# BASIC STATISTICS

## A CONCEPTUAL APPROACH FOR BEGINNERS

**Ebert L. Miller, Ed.D.**

Professor of Psychology-Education
Director of School Psychology I Programs
Ball State University
Muncie, Indiana

ACCELERATED DEVELOPMENT INC.
Publishers
Muncie, Indiana

# BASIC STATISTICS

## A CONCEPTUAL APPROACH FOR BEGINNERS

Library of Congress Number: 86-70659

International Standard Book Number: 0-915202-58-1



Technical Development: Virginia Cooper
Tanya Dalton
Judy McWilliams
Sheila Sheward

Graphic Artist: Cheryl King

**ACCELERATED DEVELOPMENT Inc., PUBLISHERS**
3400 Kilgore Avenue, Muncie, Indiana 47304
(317) 284-7511

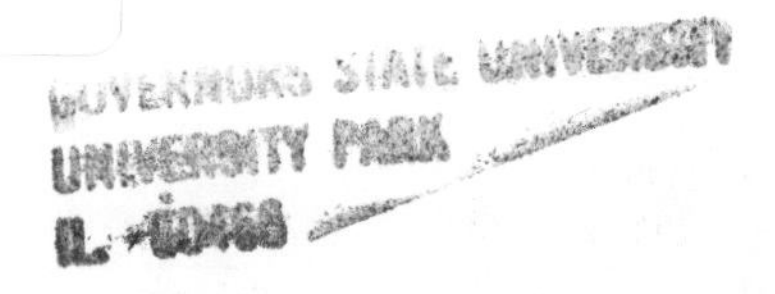

# ACKNOWLEDGMENTS

The author wishes to acknowledge the many students over the years who have motivated him to generate instructional materials that they could really understand and apply. Without the challenge of the many student questions and puzzled looks, the author would not have persisted in the time consuming task of completing this book. Throughout the long gestation period of the book, the encouragement and forebearance of his wife, Edith, has been invaluable.

# CONTENTS

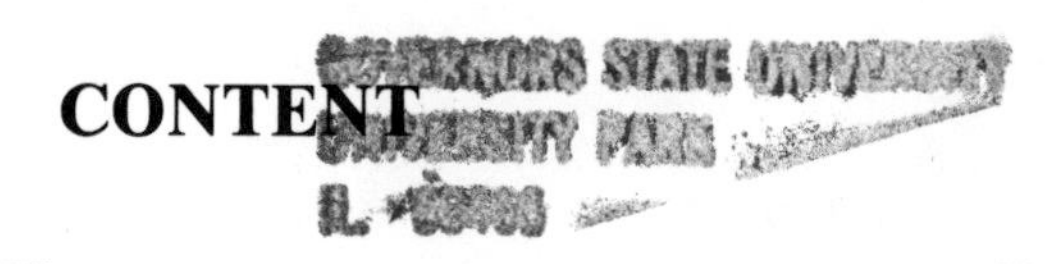



## PART 1 UNDERSTANDING STATISTICAL CONCEPTS

# PART 2 CONCEPTUALLY APPLYING STATISTICS

# LIST OF FIGURES

# LIST OF TABLES

# INTRODUCTION

This material is designed for students in upper level undergraduate or introductory level graduate courses which deal with topics in the behavioral sciences such as tests and measurements or research methodology. This material also has been helpful as an advanced organizer for students preparing to take a series of more traditional statistics courses. Three key assumptions have influenced the design of this material

First, it is assumed that most individuals studying this material have only a fundamental understanding of arithmetic.

Second, it is assumed that the amount of study time and/or instructional time to be devoted to statistical concepts will be limited. Although a chapter concerning statistics is usually a part of most tests and measurement and research methodology textbooks and such books assume prior statistics knowledge, neither capsulated versions of computational statistical approaches nor verification of assumed prior statistical knowledge receive much instructional time. As a result many students fail to achieve an adequate understanding of this key area of knowledge.

Third, it is assumed that the structure of instructional material is associated with making the material meaningful. Learning theorists generalize that meaningful material is important for the three major aspects of learning. Meaningful material facilitates initial learning of material (acquisition). Meaningful material also facilitates remembering material (retention) and further, it facilitates applying material in new and somewhat unique situations (transfer).

Because of these assumptions basic structure and sequence of the material are emphasized. Of most importance, *conceptual understandings have been emphasized in this material and formulas and computation have been deemphasized.* When time is limited and students have limited mathematical sophistication, emphasis on computation tends to result in "cookbooking" in order to get the "correct answer." When "cookbooking" occurs conceptual understanding tends to "vanish into thin air." Although it is highly desirable to develop both conceptual understanding and computational skills, this conceptual material is designed to help beginners.

This approach is not intended to provide an "all you ever wanted to know" approach to statistics. Students who plan to become serious scholars or sophisticated practitioners in any of the behavioral sciences will need additional study of statistics. Although the discerning beginner may sometimes be a bit frustrated by some of the overly simplified conceptual explanations included in this material, this is not necessarily all bad. *Remember* this is only a beginning step toward becoming a competent interpreter of statistical information.

Although "stark fear" may be an overstatement, many beginning students will readily admit to a fair amount of trepidation when the word "statistics" is mentioned. However, if a student takes enough time to study, remembers that learning new and unfamiliar material requires repetition, and focuses on how the material is applied to such things as tests, measurements, and various kinds of research methodology, the initial trepidation should soon begin to fade.

Hoping that all students will enjoy these materials is undoubtedly unrealistic, but the hope is sincere that these materials will help students to take a long step toward understanding and appreciating statistics.

The *basic goal of this material is to help students become more competent consumers.* The behavioral sciences emphasize empirical research finding. Empirical research is based on observed facts and therefore requires the use of a wide variety of tests and measurement devices. Evaluation of the effectiveness of behavior analysis techniques as well as many other intervention strategies always involves dealing with numerical data. Understanding empirical research evaluation findings and the technical aspects of interpreting tests and measurements is therefore dependent upon a basic mastery of the essentials of the language of numbers (statistics).

Some problems and exercises are interspersed throughout the material to aid the learner. These aids are set off from the basic materials as follows:

> **#1.**
>
> Understanding the structure of material to be learned makes that material more meaningful.
>
> Meaningful material is easier to ______________________,
>
> to ______________________, and
>
> to ______________________.
>
> See Appendix A for the answers.

Two recurrent ideas (concepts) are key to the structure of statistics. One is the number *two* (2) and the other is the arithmetic process of *averaging*.

Conventional courses in statistics are divided into two parts. They begin with *descriptive* statistics and then progress to *inferential statistics*. Numerical data are classified under two kinds: *continuous data* and *discrete data*. Statistics are used in two ways: one is to describe *differences* and the second is to describe *similarities* (correlation).

The concept of an arithmetic average is a key feature of statistics and each of the following terms (which will be explained in detail in later sections of this material) is an average: The *mean score* is the arithmetic average (sum all the scores and divide by number of scores). The *standard deviation,* the *variance,* the *standard error* of measurement, and other *standard errors* are also averages. As you progress through the following material keep these recurrent structural themes in mind.

The organization of the material in this book was influenced by two different studies. The first study involved a content analysis of existing statistics textbooks and statistical chapters in tests and measurement and research methodology textbooks. This analysis focused on both the statistical content and on the order in which the content was presented. The second study involved a survey of professors who teach statistics courses, tests and measurements courses, and research courses. Again the focus was on what statistical content was most relevant and on the preferred order of presentation of the statistical content. The seven modules that follow reflect the results of these two studies.

The division of the material into separate modules should facilitate use of this book. Instructors may, if they wish, skip certain modules or may reorder the sequence of these modules to fit the specific needs of their students.

Module VII is presented in a simplified scrambled text programmed format so as to aid the learner in synthesizing the material learned into concepts useful in new and somewhat unique situations (transfer of learning).

# Part 1
# UNDERSTANDING STATISTICAL CONCEPTS

MODULE I

# DEFINITION OF STATISTICS

Most people have a fair mastery of at least two languages. The first is the person's native language (e.g., English language) and the second is the language of numbers. Perhaps you never considered numbers to be a language, but in fact it is a symbolic way of communicating that has some advantages over the English language. The structure of the language of numbers is more consistent than the structure of the English language. Rules and relationships of numbers tend to be more consistent than rules and relationships of words. Of course languages composed of words may have greater overall flexibility, but the language of numbers also has considerable flexibility.

The important idea is that numbers are a type of language that you already understand fairly well. Statistics is not really completely new to you and a little new terminology should not be terribly frightening.

Definitions of statistics may differ somewhat from author to author. The definition used in this material emphasizes that statistics should be considered to be a useful *tool* and further specifies what that tool helps you do.

✓*Statistics is a tool for organizing, describing, and inferring from collections of numerical evidence.*

If you have ever tried to drive a nail with the heel of your shoe, you probably found that both the process and the result left a lot to be desired. A hammer, the properly designed tool for driving a nail, certainly would have made the job easier and the result more dependable.

As the definition points out, statistics is the correct tool to help us organize numerical data so that we can communicate (describe) accurately and clearly relevant aspects of those data. In some instances we also may draw inferences about the probable significance of observed similarities or differences.

**#2.**

Statistics should be viewed as the correct __________

to enable a person to ____________________,

to ______________________________,

and sometimes to draw ____________________

from collections of numerical evidence (data).

See Appendix A for the answers.

## ORGANIZING DATA

✓ In order to use statistics, observations about characteristics of people, places, times, and things must be gathered and converted into numerical form. These characteristics (e.g., weight, height, age, intelligence test scores, and sex) are called *variables.* Observations of a number of cases with respect to a variable yields evidence of individual differences of either magnitude or kind.

✓ Numerical evidence showing differences in magnitude (size, volume, or amount) is expressed as *measurement data.* If individual cases differ in a way that may be expressed in units along a continuum, then measurement is used (i.e., inches in height, pounds in weight, or points of I.Q.). Variables that can be measured are said to be *continuous variables.* Numerical data showing differences in magnitude are entitled either *measurement data* or *continuous data,* both meaning the same thing. A numerical continuum can be thought of as the horizontal base line of a graph. A continuum has zero or a small number at the far left end and as the units progress toward the far right end the numbers increase in size. An analogy would be a yard stick laid flat on a horizontal surface with zero at the left end and 36 inches at the far right end. The units of measurement progress in inches to the far right end. The units of measurement progress in inches (or fractions of an inch) from zero to 36.

Numerical evidence showing differences in kind (e.g., male, female, marital status, number of dropouts during a particular semester) is expressed as *enumerations.* The number of cases which fall into each independent, often referred to by the technical term of *discrete,* category are counted (enumerated), summed, and constitute *enumeration data.* Numerical data showing differences in kind are called by either of two terms, *enumeration data* or *discrete data.* Although a distinction can be made between quantitative and qualitative enumeration data, this distinction is not crucial for the task at hand and will be left for your future education.

Further organization of data depends on what the organizer wishes to communicate. If the results of how a sixth grade class scored on a group intelligence test is to be presented in a table, then making an array of those scores (putting them in rank order that is either increasing or decreasing size of scores) is a logical step. Whether to report every possible score between the lowest and highest scores made by members of the class, or to group scores (e.g., 100-104, 105-109, and so forth) is an arbitrary decision. How much detail do you wish to retain in your description? How long will the table be if you do not group the scores? Do you wish to compute into 4 equal segments (quartiles), 10 equal segments (deciles), or 100 equal segments (percentiles)? Do you wish to construct a cumulative growth curve (ogive)? Questions like these hopefully influence your decision so that the resultant table will clearly and concisely communicate adequate information to the person who studies it.

If you wish to show how well the distribution of scores (for a norming sample) of a newly created I.Q. test approximate a ''normal'' or symmetrical distribution, then you should prepare a graphic representation of those scores.

Although the person organizing data makes some arbitrary decisions about what to present and how to present it, these decisions should facilitate clear and relevant descriptions of available data.

## DESCRIBING DATA

### Tabular Presentations

One important way of summarizing numerical evidence is to present it in table form.

✓Tables in which the entries are frequencies (number of cases) resulting from a two-way classification of characteristics are called *contingency tables.* The *enumeration data* (discrete data) organized in Table 1 describe the marital status of male and female subjects of a research study.

Table 1. Contingency Table of Marital Status of Subjects

| | Male | Female | Both |
|---|---|---|---|
| Total Number | 69 | 58 | 127 |
| First Marriage, Lived with Spouse | 37 | 22 | 59 |
| Married, Separated | 2 | 1 | 3 |
| Spouse Deceased, Remarried | 3 | 2 | 5 |
| Spouse Deceased, Not Remarried | 2 | 5 | 7 |
| Divorced, Not Remarried | 4 | 9 | 13 |
| Divorced, Once, Remarried | 6 | 6 | 12 |
| Divorced More Than Once, Remarried | 4 | 1 | 5 |
| Never Married | 11 | 12 | 23 |

## Measurement Data

*Often measurement data* (continuous data) needs to be presented in table form. When this is done, it is called a frequency distribution. Such a table contains the frequency (f = number of times) with which members of the group being studied achieve a particular score or a particular level of measurement on the variable under consideration. Table 2 contains a frequency distribution of scores on a social studies achievement test. In this instance, the scores have been arbitrarily grouped to shorten the table.

Table 2. Frequency Distribution of Scores on a Social Studies Test

| Scores | Frequency (f) | Cumulative Frequency from bottom (cum. f) |
|---|---|---|
| 74-74 | 1 | 77 |
| 65-69 | 4 | 76 |
| 60-64 | 7 | 72 |
| 55-59 | 8 | 65 |
| 50-54 | 12 | 57 |
| 45-49 | 15 | 45 |
| 40-44 | 10 | 30 |
| 35-39 | 9 | 20 |
| 30-34 | 6 | 11 |
| 25-29 | 3 | 5 |
| 20-24 | 2 | 2 |

A column for the cumulative frequency from the bottom of the distribution of scores has been included in this table. Including this information facilitates constructing a cumulative frequency graph (ogive) or computing partition values such as the median or percentile ranks. If the ogive or partition values were not relevant, then the cumulative frequency data would not be included.

#3.

A contingency table summarizes ______________ data.

A frequency distribution summarizes ______________ data.

See Appendix A for the answers.

The scores in Table 2 are grouped into intervals of five measurement units. If the data has not been grouped, the table would have been an extremely long one. However, this makes necessary the consideration of the actual (real) limits (see Note 1) of continuous data intervals.

If someone asks your weight, you probably would report that weight to the nearest pound. But a reported weight of 150 pounds would actually indicate a weight somewhere between 149.5 pounds and 150.5 pounds. So 150 pounds is the reported value and 149.5 pounds and 150.5 pounds are the lower and upper actual limits of the interval.

In Table 2 the bottom reported interval of scores is 20-24. The lower actual limit of this interval is 19.5 and the upper actual limit is 24.5. Figure 1 shows that the reported score interval of 20-24 contains five units of measurement ranging along the continuum from 19.5 to 24.5.

When grouping scores, such as those in Table 2, an odd number of units (5) results in the midpoint of the interval being a whole number (22) rather than a fraction. Because the midpoint of each interval will be the value assigned to all cases within the interval for any further computation, for convenience choose an internal so as to have the midpoint a whole number.

Using the midpoint to represent all cases within the interval is not purely arbitrary. If this procedure is used consistently, then logically one can expect that some of the scores will be lower than the assigned midpoint and some will be higher. Assigning the midpoint will probably tend

---

**Note 1:** Different authors may label these limits as actual (real) limits while other authors may label them as theoretical limits. Despite the semantic confusion both labels indicate the same limits.

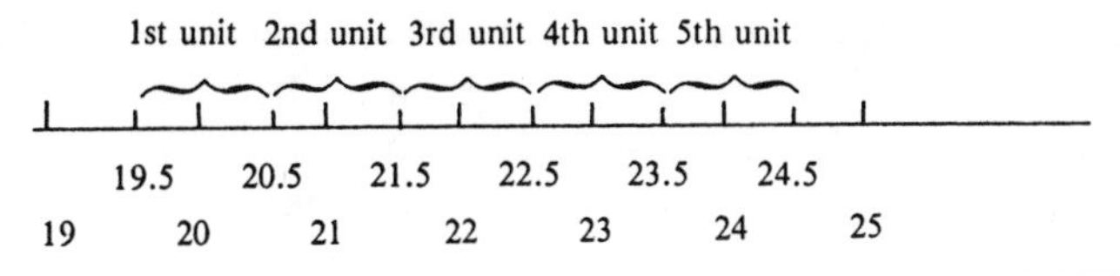


Figure 1. Continuum showing the range of scores for the reported interval of 20-24.

to average out the inaccuracy of the assigned scores as compared with the real scores. The ideas of *averaging* and what will *probably happen* (probability) are key concepts for making inferences about descriptions of collections of numerical evidence.

The idea of upper and lower actual (or if you prefer theoretical) limits of a measurement unit can present a problem when trying to compute a value such as the median (one of the measures of central tendency to be studied later). To compute the precise point between the lower and upper limits of a measurement unit so that exactly one half of the cases fall above that point and exactly one half of the cases fall below that point may be difficult to reason out so a computational formula may become a crutch. This use of a formula is appropriate so long as the user conceptually understands what has been computed. If conceptual understanding is lost, then "cookbooking" has taken over.

**Graphic Presentations**

Graphs are a kind of pictorial representation that give insight into the nature of the measured characteristic for the entire group studied. A number of graphic forms are used in statistics. Those data summarized in Table 2 will be used to demonstrate those graphic forms most often used in measurement and evaluation (i.e., a *histogram,* a *frequency polygon,* and a *smoothed frequency distribution*).

Before you become overly impressed or depressed with this terminology, you should remember that you first learned about graphs in third or fourth grade arithmetic. A *histogram* is really a bar graph, a *frequency polygon* is really a *line graph* and the *smoothed frequency polygon* is really an adjusted (smoothed) *line graph.*

Each graph has a vertical axis (*ordinate*) and a horizontal axis (*abscissa*). These labels name the two lines but do not explain their functions. The *ordinate* has the frequency (number of cases) running up the vertical axis. The *abscissa* is really the base continuum and has the scores (measurements) running from low scores on the left to higher scores on the right. In the following example, frequency ranges along the vertical axis and the social studies test scores range along the horizontal axis.

Figure 2 is a histogram (bargraph) constructed on graph paper to show how easily it can be constructed by drawing a horizontal line across each interval (actual limits of intervals shown) for the social studies test scores in Table 2 at the level corresponding to the frequency within the interval.

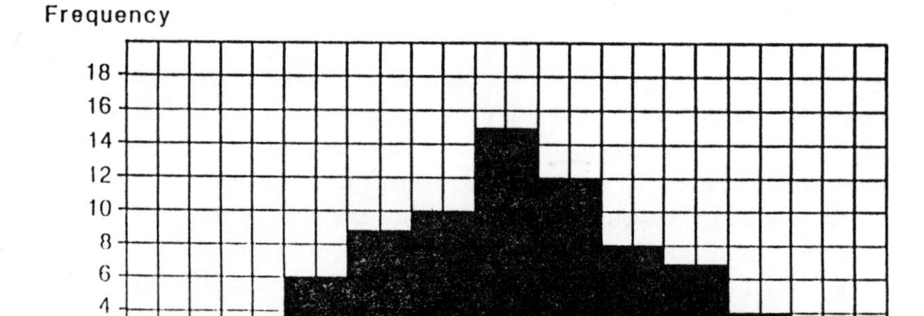


Figure 2. Histogram.

Figure 3 is a frequency polygon (line graph) constructed by joining the plotted mid-points (the level corresponding to the frequency within the interval) of each interval.

Figure 4 is a smoothed frequency polygon constructed by joining the plotted mid-points of each interval. However, these mid-points must be computed by an averaging process. The mid-point of an interval is computed by adding the total of the frequencies in that interval and the two adjoining intervals (the one above and the one below) and dividing this total by three (the number of intervals). For example, the mid-point of the interval 25 to 29 is computed by adding the frequencies 3, 6, and 2 and dividing the result of 11 by 3. Thus it is determined that the point to be plotted for the interval (25 to 29) is 3.67. For the interval 20 to 24, the frequencies 2, 3, and 0 are summed and the result 5 is divided by 3. Thus, the point to plot for this interval is 1.67. When this procedure is carried out for every interval the peaks and valleys of the frequency polygon are smoothed out.

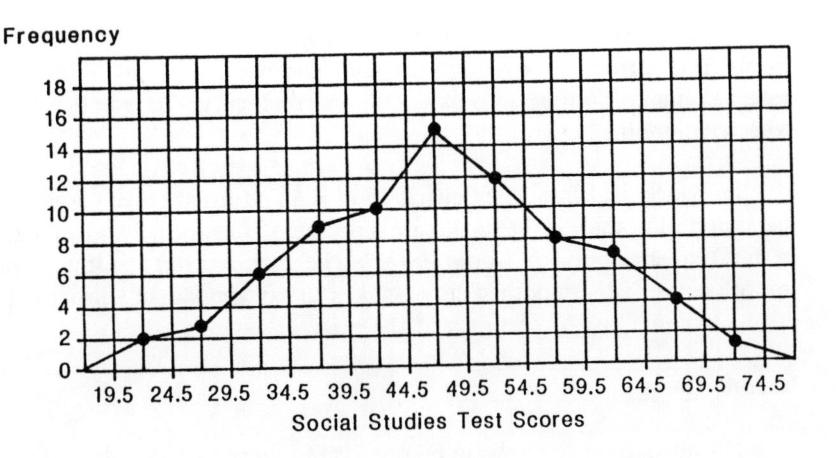


Figure 3. Frequency polygon.

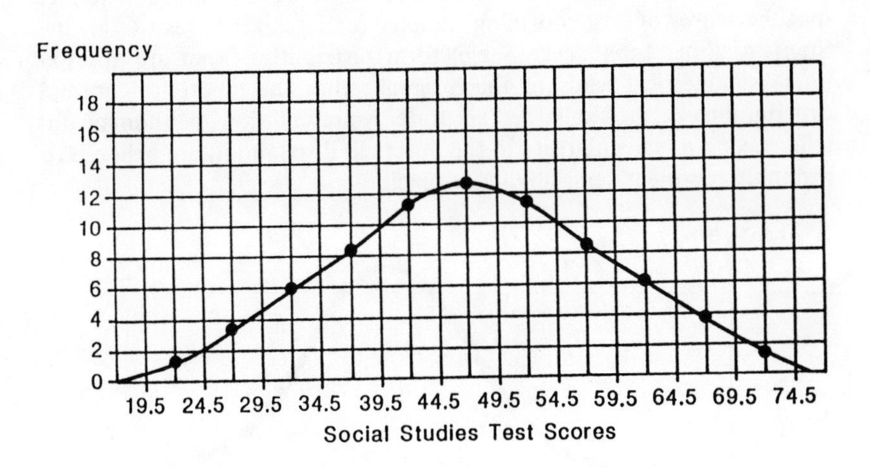


Figure 4. Smoothed frequency distribution.

Graphic representation of a frequency distribution is made so that one may visually inspect the shape of the distribution to determine if it approximates a normal (symmetrical) curve distribution or some other expected distribution.

Although the histogram and the frequency polygon are reasonably simple to construct, they require the reader to use some imagination in order to make comparisons with the normal curve or some other expected smoothed curve.

The smoothed frequency distribution is an adjusted curve based on the assumption that, if the data were more numerous, peaks and valleys in the frequency polygon would *probably* be *averaged out* resulting in a smooth curve. This smoothed curve makes visual comparisons easier but results in some loss of accuracy of the original data.

**Normal Curve**

The normal curve shown in Figure 5 is familiar to many people. This symmetrical, bell shaped curve is quite common and is a graphic representation of the theoretical distribution of many human characteristics (e.g., intelligence). The assumption is that if we could measure the intelligence of the infinite population of human beings (everyone who ever lived, now lives, and will live in the future), those scores would form a normal distribution curve when graphed. The fact that the scores of large norming samples for intelligence tests (e.g., the Stanford Binet Intelligence Scale) form distributions that are approximate symmetrical leads to the inference that the theoretical normal distribution is probably an accurate representation of human intelligence. Any irregularities in the observed distribution are believed to probably represent sampling inadequacies.

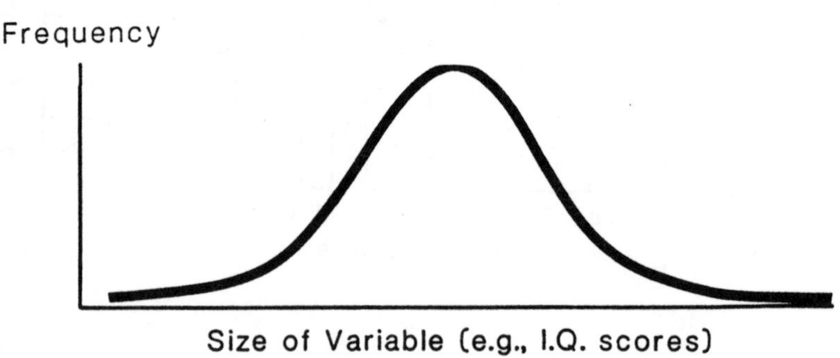


Figure 5. Normal curve.

Because graphic representations take considerable space in a publication you may simply be told in a few words about the shape of the distribution. From such a written description, you are expected to be able to visualize what the distribution looks like. For example, you might read that administering a new I.Q. test to a norming sample resulted in a distribution of scores that formed a "negatively skewed leptokurtic curve." Obviously you need to understand some new terminology.

Distribution curves have two characteristics, *skewness* and *kurtosis.*

If a distribution is not *symmetrical,* then it is said to be *skewed.* When the "tail" of the skewed (twisted off being symmetrical) distribution points to the left (See Figure 6), the distribution is said to be negatively skewed.

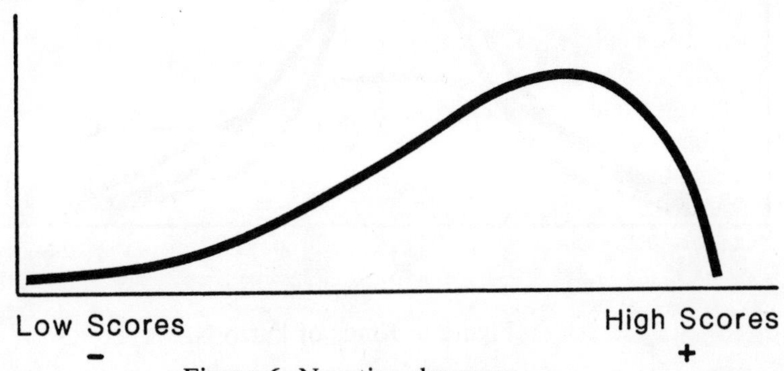


Figure 6. Negative skewness.

In Figure 7 is shown a distribution that is positively skewed with the tail pointing to the right end of the scale.

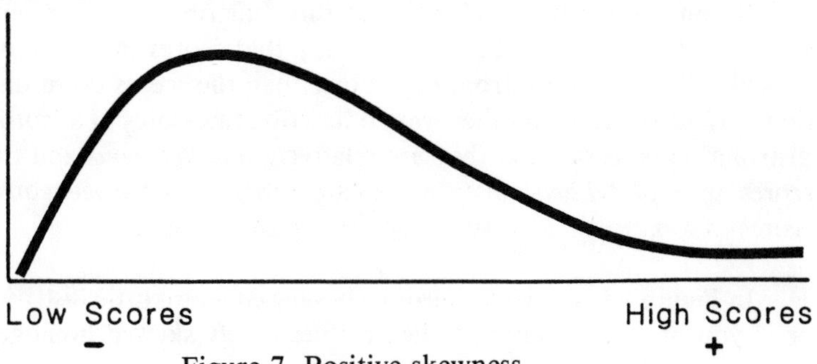


Figure 7. Positive skewness.

Although the labels negative skewness and positive skewness may seem illogical, the logic of these labels will be explained in the following section on measures of central tendency.

*Kurtosis* indicates the degree of peakedness of a curve. In Figure 8 is shown different kinds of kurtosis. Curve A is *leptokurtic,* Curve B is *mesokurtic,* and Curve C is *platykurtic.*

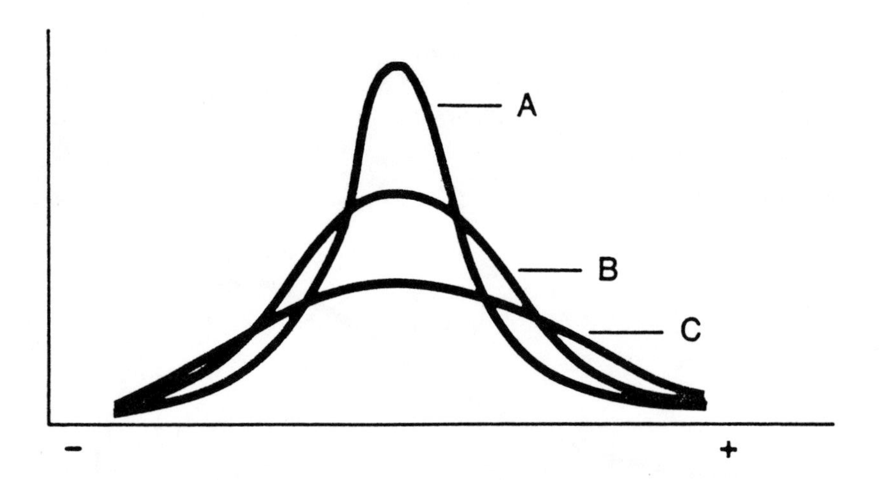


Figure 8. Kinds of kurtosis.

If you assume that each of these three distributions represents the same number of cases, then you can say that scores in distribution C (*platykurtic*) vary more from each other than the scores do in distribution A (*leptokurtic*). Another way to describe the scores in a leptokurtic distribution is to say that they are relatively *homogeneous* and that the scores in a *platykurtic* distribution are relatively *heterogeneous.* The scores in a *mesokurtic* distribution are typical (normal).

In Figure 9 is shown a "negatively skewed leptokurtic distribution" or if you prefer, it may be called a "negatively skewed homogeneous distribution."

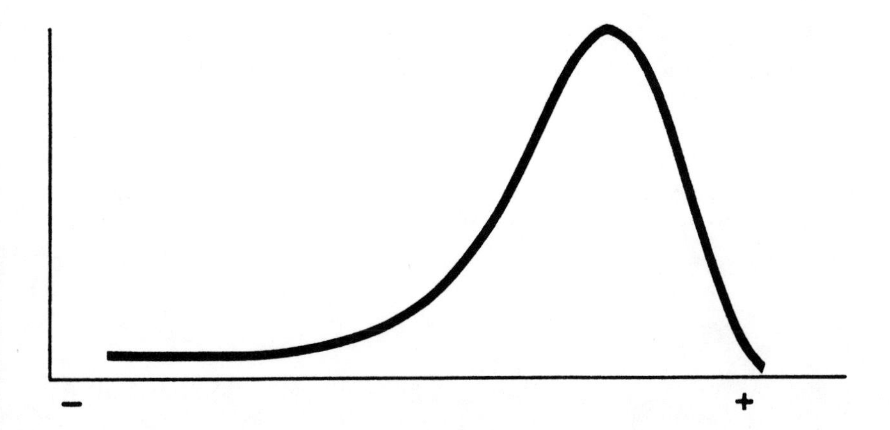


Figure 9. Negatively skewed leptokurtic curve.

# MEASURES OF CENTRAL TENDENCY AND VARIABILITY

In order to understand the implications that shapes of a distribution have for interpreting scores and their accuracy you must learn about *central tendency and variability.*

## CENTRAL TENDENCY

Suppose a person is asked to describe, using just one number, how well a group scored on a particular test. Where would the most representative figure fall in most distributions? It would tend to be in the center (*central tendency*) of the distribution.

The three measures of central tendency you need to consider, in order of their general importance, are the *mean,* the *median,* and the *mode.*

## Mean

The *mean* is the arithmetic average of all the scores in a distribution. It is computed by summing all the scores and dividing by the number of scores. Because the size (magnitude) of each score and the number of scores in the distribution influence the final result, the mean is the most mathematically sensitive of the measures of central tendency. The mean is often likened to the balance point (center of gravity) of a distribution. Therefore, *the mean is the most appropriate for use unless the distribution of scores is extremely skewed* (asymmetrical).

An arithmetic achievement test is given near the end of a school year to a large representative sample of third grade children in a large city school system. This test is a measurement or norm referenced test. The test was designed so that no one would miss all the questions and no one would answer all of the questions correctly. Norms for a national sample are available, but the local school system wishes to develop norms for the local situation.

The local norms will enable the local school system to better determine the effects of some new curricular changes which they plan to try. What would probably be the best measure of central tendency to compute for the local sample? They would add (sum) all the scores of the third grade children who took the test and then they would divide this sum by the number of scores for these children. This would provide the mean score for the group (local norm group). This mean score could then be compared to the national mean score or with the mean score of a later representative group who have been instructed with the new curricular modifications. In this way mean gains or losses can be computed and inferential statistical procedures can be computed to estimate whether noted gains or losses are probably related to the curricular changes.

## Median

The *median* is essentially a partition value. It is the point on the base continuum where raising a vertical partition will divide the distribution in half. With ungrouped data the median is the middle score (median score) and is relatively easy to compute. Divide the number of scores by two then count up (or down) the rank ordered scores to locate the middle score. If the number of scores is an even number, the median is the value midway between the two middle scores.

With grouped data (scores grouped into intervals) computation requires determining in which interval the median lies. Then the precise distance the median lies above the bottom (lower actual limit) of the interval is computed. This distance is added to the lower actual limit of the interval and the result is the median (Note 2).

The median does not consider the size (magnitude) of each score other than to place it in the top or bottom half of the scores. *Where the distribution of a variable is highly skewed, the median may be the most representative measure of central tendency to describe what is typical.* For example, the distribution of wealth in our country is relatively skewed and the median national-income is more often reported than is the mean income.

The salary committee of a school district wishes to compare salaries of teachers and administrators with teachers and administrators in other districts, as well as making comparisons with various other occupational groups (e.g., truck drivers). Because the distribution of wealth (and probably salaries) tends to be skewed, the median salaries will probably be chosen for the comparisons. The salary that represents the demarcation between the highest fifty percent (50%) of the salaries and the lowest fifty percent (50%) of the salaries is the median salary.

The median is the partition value which is a measure of central tendency. Other partition values often used in tests and measurements are *quartiles, deciles,* and *percentiles.*

## Quartiles

*Quartiles* are the points on the base continuum where raising partitions will divide the distribution into four equal parts (twenty-five percent of the scores in each quarter). $Q_1$ is the first quartile (one-fourth of

---

**Note 2:** The reader may check a conventional statistics text book for examples of computation and the formula for computing the median. However, you should be warned that this is the point in statistics where computation may begin to seriously compete with conceptual understanding. If your algebra background is old, weak, or non-existent, formulas may become crutches to help you obtain numbers which you do not really understand.

the scores below and three-fourths of the scores above) and $Q_3$ is the third quartile (three-fourths of the scores below and one-fourth of the scores above). The median is the second quartile with one-half of the scores above and one-half of the scores below. Of course one-fourth of the scores fall between $Q_1$ and the median and one-fourth of the scores fall between the median and $Q_3$.

## Deciles

*Deciles* are *points* on the base continuum where raising partitions will divide the distribution into ten equal parts (e.g., 10% of the scores fall between the 1st decile and the 2nd decile).

## Percentiles

*Percentiles* are *points* on the base continuum where raising partitions will divide the distribution into one hundred equal parts (e.g., one percent of the scores fall between the 98th percentile and the 99th percentile). Percentiles are often abbreviated by use of the percent sign and additions of letters ile, e.g., 98%ile.

*Quartiles, deciles,* and *percentiles* will be discussed further in a later section concerning test interpretation.

## Mode

The *mode* is the most easily computed (recognized) measure of central tendency. It is the score which occurs most frequently in the distribution. *Where the scores have been grouped, the mid-point of the interval containing the largest frequence is usually designated as the mode.*

The mode is the least used measure of central tendency for making comparisons between different groups. *The mode is most often used to describe unusual distributions.* Distribution of scores on classroom achievement tests for undergraduate education students taking summer school classes some times takes the form of a bimodal distribution (See Figure 10). This unusual distribution of scores for summer school students as compared with the distribution of scores for academic year students may be the result of several factors. The summer school undergraduate students may represent two quite different groups of students. For example, one group chooses to attend summer school because they have done poorly and wish to attempt to better their previous scores and grades. The other group represents students who are high achievers in class and who also are involved in many co-curricular activities. They choose to attend summer school in order to have lighter academic year class loads so that they may concentrate more time on co-curricular activities. The poor achievers do not achieve very well during the summer and the high achievers continue to achieve very well. The result is a bimodal distribution of scores that is quite dissimilar to the academic year distribution of scores.

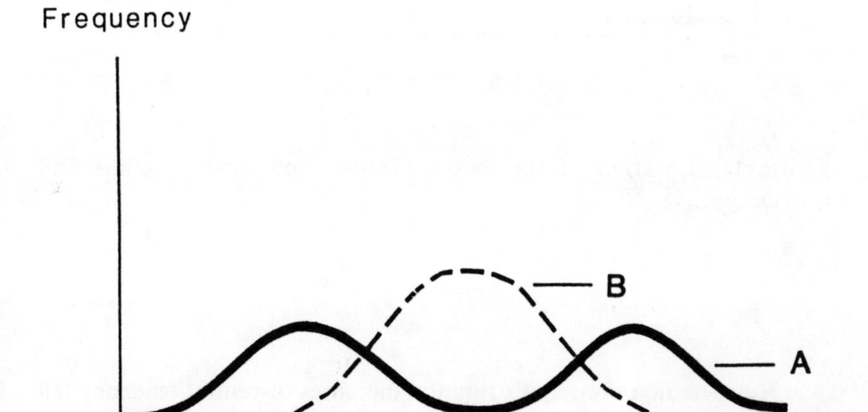


Figure 10. Distribution of scores for summer school students (A) and academic year students (B).

### Measures of Central Tendency and Distribution Shape

In a symmetrical distribution (e.g., the normal distribution) the mean, the median and the mode fall at exactly the same place on the base continuum (see Point A in Figure 11).

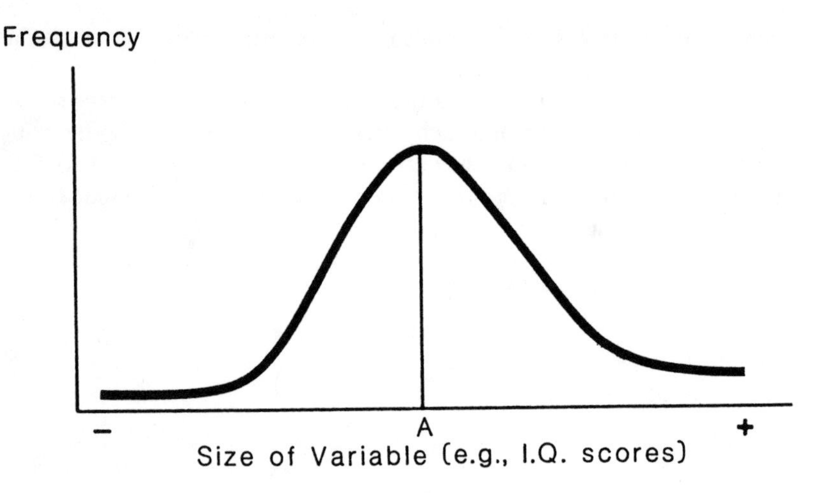


Figure 11. Location of the mean, median, and mode—symmetrical distribution.

However in a skewed distribution measures of central tendency fall at different points along the base continuum. In Figure 12 you can see that the mean has been pulled (displaced) toward the right or positive end of the scale (Note 3). In a negatively skewed distribution the mean would be displaced toward the left or negative end of the scale. This displacement of the mean toward the atypical scores (tail of the distribution) is the reason for the positively skewed or negatively skewed labels. Distribution of wealth often forms a positively skewed distribution and scores on a mastery test often form a negatively skewed distribution.

---

**Note 3:** Scores to the right of (higher than) the measure of central tendency are designated as positive (+) and scores to the left of (lower than) the measure of central tendency are designated as negative (-). These designated signs (+ or -) should not be interpreted as value judgments, but simply as an indication of whether a particular score is above or below the measure of central tendency. For example, a high score on a test that measures "dishonesty" would not be viewed as positive (from a value judgment point of view) by most people. However, a high intelligence test score would probably be viewed as positive (from a value judgment point of view) by most people.

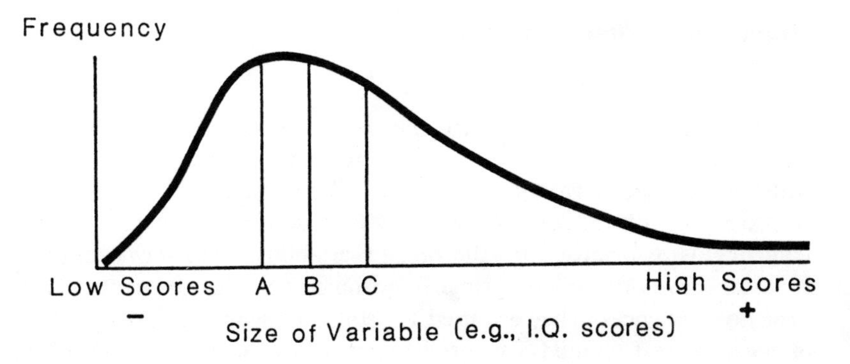


Figure 12. Positively skewed distribution—mode (A), median (B), and mean (C).

**#4.**

| A | | | B | | |
|---|---|---|---|---|---|
| f* | X** | fX*** | f* | X** | fX*** |
| 1 | 8 | 8 | 1 | 8 | 8 |
| 2 | 7 | 14 | 3 | 7 | 21 |
| 3 | 6 | 18 | 3 | 6 | 18 |
| 2 | 5 | 10 | 2 | 5 | 10 |
| 1 | 4 | 4 | 1 | 1 | 1 |
| 9 | | 54 | 10 | | 58 |

* f is frequency
** X is raw score
*** fX is f times X

Compute the mean score for each distribution

Mean for A is ____________.

Mean for B is ____________.

Distribution A is ______________________ skewed.

Distribution B is ______________________ skewed.

See Appendix A for the answers.

**Measures of Central Tendency and Variability**

Each measure of central tendency is usually associated with a different measure of variability. The *mean* (central tendency) is associated with the *standard deviation* (variability). The *median* (central tendency) is associated with the *quartile deviation* (variability). The *mode* (central tendency) is associated with the *range* (variability). The *mean* and the *standard deviation* are of prime importance for measurement in the behavioral sciences. They are used for standard scores, standard errors of measurement, standard errors of estimates, and various inferential statistical procedures used to test hypotheses.

## VARIABILITY

In addition to characterizing a distribution in terms of one point along a continuum, descriptive values can be used to describe the spread of scores. Variability is the amount of spread of scores from a measure of central tendency. When the scores are bunched and the spread is small the variability is small (e.g., a leptokurtic or homogeneous distribution). When the scores are more widely dispersed the variability is large (e.g., a platykurtic or heterogeneous distribution).

An example of how the variability of scores may be thought to affect practice is contained in the rationale for ability grouping of students. The belief is that if the variability in ability of a particular class could be made quite small (based on such things as I.Q. scores, achievement test scores, and teacher ratings), then one curriculum would be appropriate for all the students in that particular class. However, if the variability of ability was quite large (as it might be in an ungrouped class), then special adaptations of the curriculum would be necessary for many of the students, resulting in multiple curricula.

**Range**

The *range* is the simplest and least meaningful measure of variability. It is the difference gotten by subtracting the lowest score in a distribution from the highest score in that distribution. Sometimes a differentiation is made between the *simple range* and the *inclusive* (absolute) *range*.

The *inclusive* (absolute) *range* is the highest score minus the lowest score plus one so that all the scores are included within the range. The one unit is added to extend the upper and lower actual limits of the range so that both the lowest and highest score are included within the range.

The range gives some idea about the overall spread of the scores but does not give any idea about the spread of the scores around the central tendency. With a skewed distribution the range generally indicates a much larger degree of variability than is typical of the distribution. For example, the last season that the author bowled his scores ranged from a low score of 135 to a high score of 258. However, over 80% of the scores ranged between 145 and 165. The range of 123 points indicates a much greater degree of variability than was typical. Unfortunately scores over 180 were exceedingly rare and the actual leptokurtic positively skewed distribution was not highly valued by any member of the bowling team.

**Quartile Deviation**

The *quartile deviation* (Q or Q.D.) is frequently called the semi-interquartile range. It is computed by subtracting the score of the first quartile ($Q_1$) from the score of the third quartile ($Q_3$) and dividing by two. The interquartile range ($Q_3—Q_1$) is the range of the middle 50% of the distribution. The *quartile deviation* (Q) is one-half of the interquartile range.

In a normal distribution the quartile deviation (Q) is the distance between the median and either the first quartile ($Q_1$) or the third quartile ($Q_3$). In a skewed distribution this is not so.

If the I.Q. score for $Q_1$ (first quartile or 25 %ile) was 89 and the I.Q. score for $Q_3$ (third quartile or 75 %ile) was 111 then Q could be computed as follows:

$$\frac{Q_3 - Q_1}{2} = \frac{111-89}{2} = \frac{22}{2} = 11$$

The range of 22 points (111—89) includes the middle 50% of the scores on that particular test and Q the quartile deviation is 11 points.

Partition values present interpretation problems that may mislead persons trying to understand scores. In Figure 13 you can see that the distance along the continuum (measurement distance) is smaller for 10% of the scores (e.g., 40 to 50%ile) near the middle of the distribution than it is for 10% of the scores at an extreme (e.g., 90 %ile to the top of the distribution).

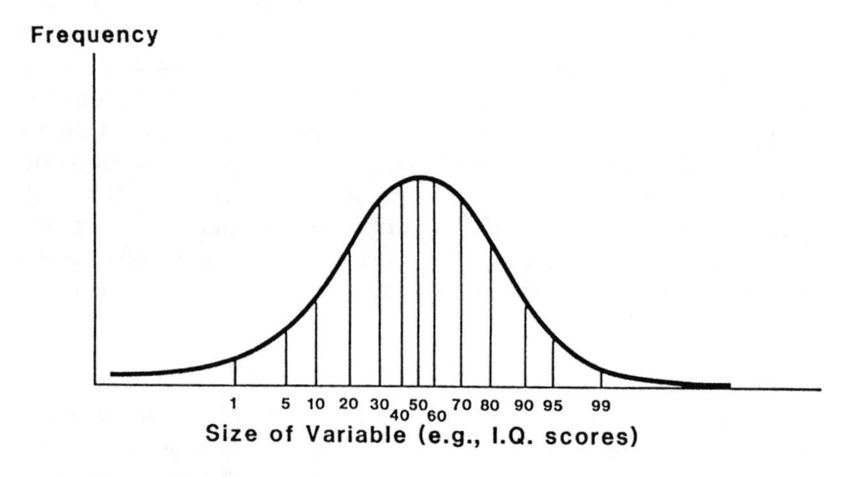


Figure 13. Percentile equivalents, normal distribution.

## Standard Deviation

Having equal intervals along the continuum would make the interpretation of the test results less subject to error. To accomplish this the raw scores of tests must be converted into standard scores. A *standard score* has a set mean and standard deviation. The *standard deviation* represents a set distance along the continuum (including standard measurement intervals within the standard deviation).

The *standard deviation* is one of the most important statistical concepts for interpreting test scores. It is a key to understanding *standard scores, normalized standard scores, standard errors of measurement, other standard errors,* and also analysis of variance inferential statistical tests.

Because the standard deviation is a complex concept, simply outlining how it is computed (as was done for the range and quartile deviation) will not provide an adequate definition.

You will remember that when you wished the one best number to represent a distribution you computed the mean (arithmetic average). To get the best number to represent the spread or variability of the scores around the mean you will compute a different kind of arithmetic average.

The *standard deviation,* in simple terms, is the *average distance that all the scores vary or deviate from the mean.* In Table 3 is shown a very simple distribution of scores (X), with a mean of 11, which can be used to demonstrate (Note 4) how the *mean deviation,* the *variance,* and the *standard deviation* are computed.

The *mean deviation* has been called an "illegitimate" statistic. A less picturesque way of labeling the mean deviation would be to say that it is a "terminal" statistic. When it is computed, the signs of the deviation scores are ignored so that it cannot properly be used to compute other measures. Although it is rarely used today, it should help you understand the meaning of standard deviation.

Table 3. Distribution of Seven Scores With a Mean Score of 11

| Scores X | Deviation From the Mean - Ignoring the Sign $\lvert x \rvert$ | Deviation From the Mean Including the sign x | Squared Deviation from the Mean $x^2$ |
|---|---|---|---|
| 14 | 3 | +3 | +9 |
| 13 | 2 | +2 | +4 |
| 12 | 1 | +1 | +1 |
| 11 | 0 | 0 | 0 |
| 10 | 1 | -1 | +1 |
| 9 | 2 | -2 | +4 |
| 8 | 3 | -3 | +9 |
| $\sum X = 77$ | $\sum \lvert x \rvert = 12$ | $\sum x = 0$ | $\sum x^2 = 28$ |

**Note 4:** Remember computation is a procedure to an end used to help you understand the concept, *standard deviation.*

The *mean* (average) *deviation* is computed (using the data in Table 3) by subtracting the mean of the distribution from every score in the distribution to obtain the deviation score (x). A deviation score is not a score for some kind of "pervert" but is simply the number of measurement units that a score differs (deviates) from the mean. Scores above the mean have a positive sign (+) and scores below the mean have a negative (-) sign. The deviation scores, ignoring the signs, are summed ( $|x|$ ) and divided by the number of scores.

$$\text{Mean deviation} = \frac{\sum |x|}{N} = \frac{12}{7} = 1.71$$

If you had not ignored the signs, negative scores would have subtracted from the positive scores. You can see that $\Sigma X$ would be 0 and the mean deviation also would be 0. The mean deviation (when signs are ignored) provides an estimate of the standard deviation and provides the foundation for understanding that the standard deviation is an average.

The *standard deviation* is a "legitimate" statistic because the signs of the deviation scores (x) are not ignored when it is computed. Therefore, it can be used to compute other measures.

To compute the standard deviation from the data in Table 3 the deviation scores are squared ($x^2$) to make all the signs positive (+). You may remember that when two numbers of the same sign are multiplied the product is positive (e.g., +3 times a +3 equals a +9 and -3 times a -3 equals a + 9*). The squared deviation scores are summed ($x^2$) and divided by the number of scores. This average of the squared deviation scores is called the variance.*

$$\text{Variance} = \frac{x^2}{N} = \frac{28}{7} = 4$$

The *standard deviation is the square root of the variance.* Another way of saying the same thing, is to say that the *standard deviation is the square root of the average of the squared individual deviations from the mean of the distribution.* So the next step is to take the square root of the average of the squared deviation scores (variance) which will yield the *standard deviation.*

Standard Deviation $= \sqrt{\frac{\sum x^2}{N}} = \sqrt{\frac{28}{7}} = \sqrt{4} = 2$

You also can see that logically a second definition of the variance is the square of the *standard deviation.* If you have the *standard deviation,* squaring that number will yield the variance.

If you carefully study the computational steps described, you should realize that the *standard deviation* is a *mean* (average). In this case the standard deviation is the mathematically legitmate *mean* (average) *distance that all of the scores vary or deviate from the mean.*

Formulas for computing the *standard deviation* found in conventional statistics books usually expand the definition formula (standard deviation $= \sqrt{\frac{x^2}{N}}$ ) to a computational formula that makes it unnecessary to find the deviation of each score from the mean. Such formulas make computation simpler but tend to hide the meaning of standard deviation, especially for persons who are not competent in algebra.

**#5.**

| Scores X | $\lvert x \rvert$ | x | $x^2$ |
|---|---|---|---|
| 5 | 2 | +2 | +4 |
| 4 | 1 | +1 | +1 |
| 3 | 0 | 0 | 0 |
| 2 | 1 | −1 | +1 |
| 1 | 2 | −2 | +4 |
| 15 | $\sum \lvert x \rvert =$ | $\sum X=0$ | $\sum x^2=$ |

Compute the mean deviation and the standard deviation for the distribution of five scores. The mean deviation is ______________. The standard deviation is

______________________________.

See Appendix A for the answers.

**Common Symbols**

The symbols commonly used for the mean, the standard deviation and the variance are different for describing *samples* and *populations*. A *population* is all of something and a *sample* is a portion of a population. Characteristics of populations are called *parameters* and characteristics of samples are called *statistics*. Aha, the label statistics now should be a bit clearer.

Table 4. Symbols for Mean, Variance, and Standard Deviation

| | Statistics (sample) | Parameter (population) |
|---|---|---|
| Mean | $\overline{X}$ | $\mu$ (Mu) |
| Standard Deviation | s | $\sigma$ (Sigma) |
| Variance | $s^2$ | $\sigma^2$ (Sigma$^2$) |

Because symbols use less printed space than words, authors of books, periodical articles and tests manuals make frequent use of these common symbols.

What you have learned to this point is generally used to describe differences between such things as scores and mean scores for different groups. Now is the time to consider similarity between two or more scores for a group of individuals. This area of statistics is called *correlation*.

# CORRELATION

You may feel that you already have a fairly good general understanding of correlation and how to interpret correlation coefficients. You have undoubtedly encountered correlations in textbooks, professional journals or test manuals. However, correlation is not as simple as it may seem and misinterpretations are common.

Because of the complex algebraic nature of correlation, computation procedures do not help the beginner understand this concepts.

## DEFINITION

*Correlation* is used to describe the relationship between two variables. The idea of the *co-relationship* or *co-variation* of the scores on two different variables for each of the individuals in a group can be put in graphic form and also can be expressed as a single value (*coefficient of correlation* often represented by the symbol r).

## Scatter Diagram

The *bivariate frequency distribution* (Note 5) sometimes called a *scatter diagram,* provides a vehicle for conceptualizing correlation. What is the relationship between the height and weight of a group of adults (e.g., a typical college class.) If you have consulted a height and weight chart you probably noted that as heights increased the ideal weights also increased. This is an example of a *positive* correlation. If the weight for each succeedingly higher height had been increasingly lower, then the correlation would have been negative (Note 6). In a *scatter diagram,* points are plotted where the scores for the two variables intersect on the graph.

When the data in Example A of Table 5 is plotted on a scatter diagram the result is a straight line (See Example A of Figure 14). This indicates a positive perfect correlation and the computed correlation coefficient (single value indicating the degree of relationship between the two variables) would be +1.00. Example B of Table 5 and the corresponding scatter diagram (See Example B of Figure 14) is a perfect negative correlation (-1.00). Example C of Table 5 and the corresponding scatter diagram (Example C of Figure 14) is a completely random relationship (0.00 correlation coefficient) or no correlation.

---

**Note 5:** Up to this point you have learned to describe the central tendency, the variability and the shape of distributions of collections of numerical evidence for a single variable. When only *one variable* is involved, the term *univariate* is used. If you wish to describe the relationship between *two variables* then the term *bivariate* is used. Procedures for dealing with *three or more variables* are called *multivariate* analyses.

**Note 6:** In univariate descriptive statistics positive (+) indicated scores above (to the right) the center of the distribution and negative (-) indicated scores below (to the left of) the center of the distribution. In correlation (bivariate) the positive (+) indicates a *direct relationship* (the two variables either increase or decrease together). The negative sign (-) indicates an *inverse relationship* (as one variable increases the other variable decreases).

Table 5. Distribution of Scores, Examples A, B, and C

| Example A | | |
|---|---|---|
| Student | Var X | Var Y |
| A | 2 | 2 |
| B | 6 | 6 |
| C | 20 | 20 |
| D | 18 | 18 |
| E | 8 | 8 |
| F | 4 | 4 |
| G | 16 | 16 |
| H | 14 | 14 |
| I | 10 | 10 |
| J | 12 | 12 |

| Example B | | |
|---|---|---|
| Student | Var X | Var Y |
| A | 2 | 20 |
| B | 4 | 18 |
| C | 10 | 12 |
| D | 6 | 16 |
| E | 20 | 2 |
| F | 16 | 6 |
| G | 12 | 10 |
| H | 14 | 8 |
| I | 18 | 4 |
| J | 8 | 14 |

| Example C | | |
|---|---|---|
| Student | Var X | Var Y |
| A | 12 | 8 |
| B | 10 | 10 |
| C | 10 | 12 |
| D | 12 | 14 |
| E | 12 | 12 |
| F | 8 | 14 |
| G | 10 | 8 |
| H | 14 | 12 |
| I | 8 | 10 |
| J | 12 | 10 |

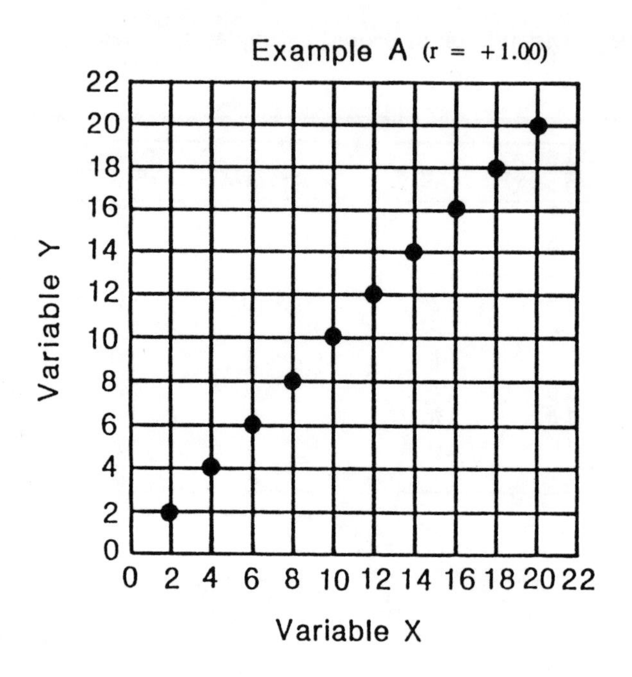


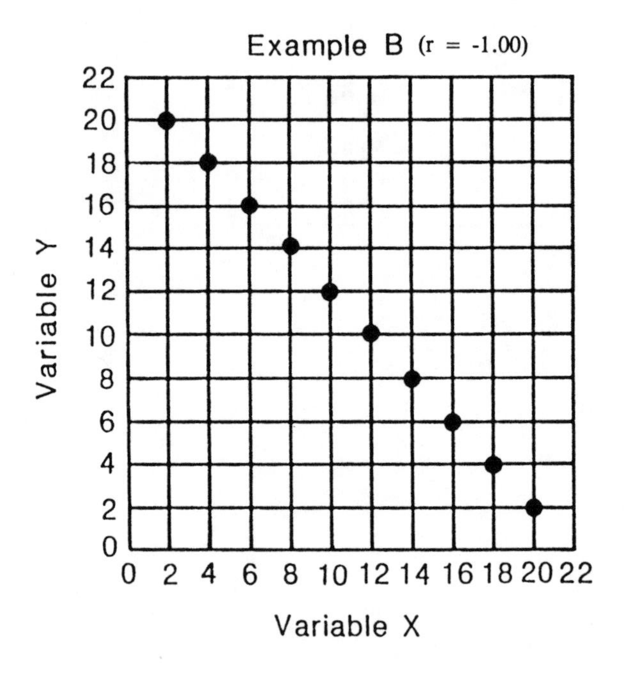


Figure 14. Scatter diagrams with plus 1, minus 1, and zero correlations.

Figure 14. Continued.

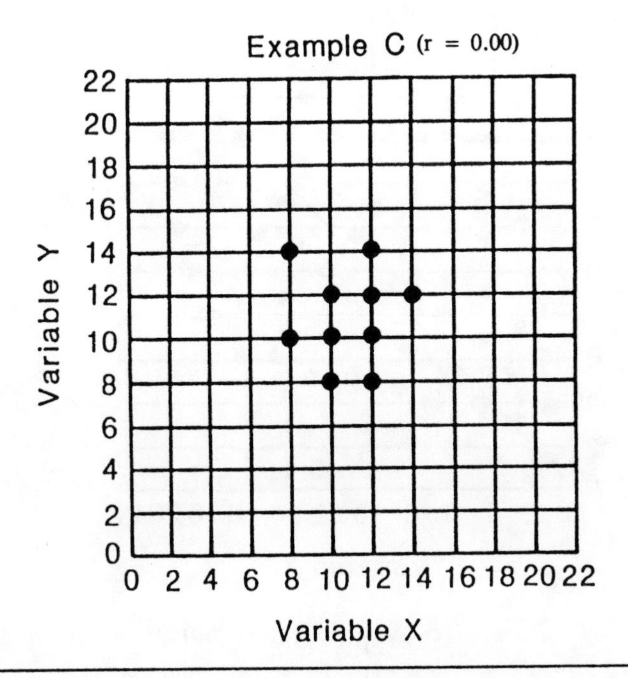


---

Coefficients of correlation are never larger than 1.00 (unity) and a coefficient of 0.00 indicates no relationship between the two variables. Coefficients between 0.00 and 1.00 (e.g., .80) indicate some relationship between the two variables with some of the plotted points deviating from the plotted line of perfect correlation. If you drew a line from each plotted point to the line of perfect correlation (so that these two lines intersected at a right angle), you would have a deviation distance score for each plotted point which did not fall exactly on the line of perfect correlation. The larger the combined deviations of all plotted points, the smaller the correlation coefficient (nearer 0.00). In Figure 15, Example A is shown a relatively larger amount of deviation of plotted points (coefficient approximately .40) than that found in Example B (coefficient approximately .80). Conceptually you should see the similarity between these deviations and the deviation scores in the preceding section on standard deviation. If you take computational statistics this association may help you to better understand the mathematical assumptions associated with correlation.

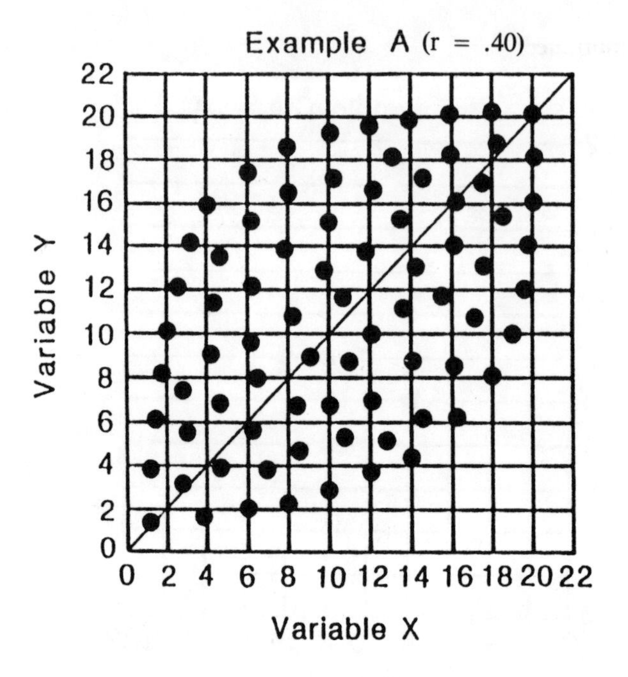


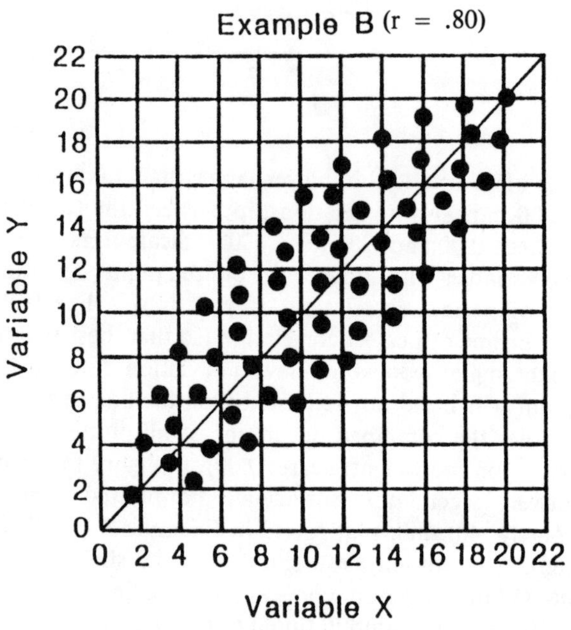


Figure 15. Scatter diagrams with correlations between zero and plus one.

**#6.**

A B

Which diagram has the correlation closest to 0.00?

(A or B) ______________________

Which diagram shows a negative (-) correlation?

(A or B) ______________________

See Appendix A for the answers.

## Venn Diagrams

Because correlation is a complex and abstract concept, another way of visualizing relative degrees of relationship is presented in Figure 16. Venn diagrams (circles) are used to show that the amount of overlap between variables X and Y (shaded area) indicates the amount of variance they have in common (i.e., the degree of correlation between X and Y). In general terms (Note 7), Example A has a high correlation, B a moderate correlation, C a small correlation, and D no correlation. One weakness of this visual approach is the inability to determine if the correlation is positive or negative. However, this approach does demonstrate co-relationship (correlation) of the variables (amount of variance they have in common).

---

**Note 7:** See the following section concerning interpretation of correlation for an explanation of why the qualifier "general terms" was used.

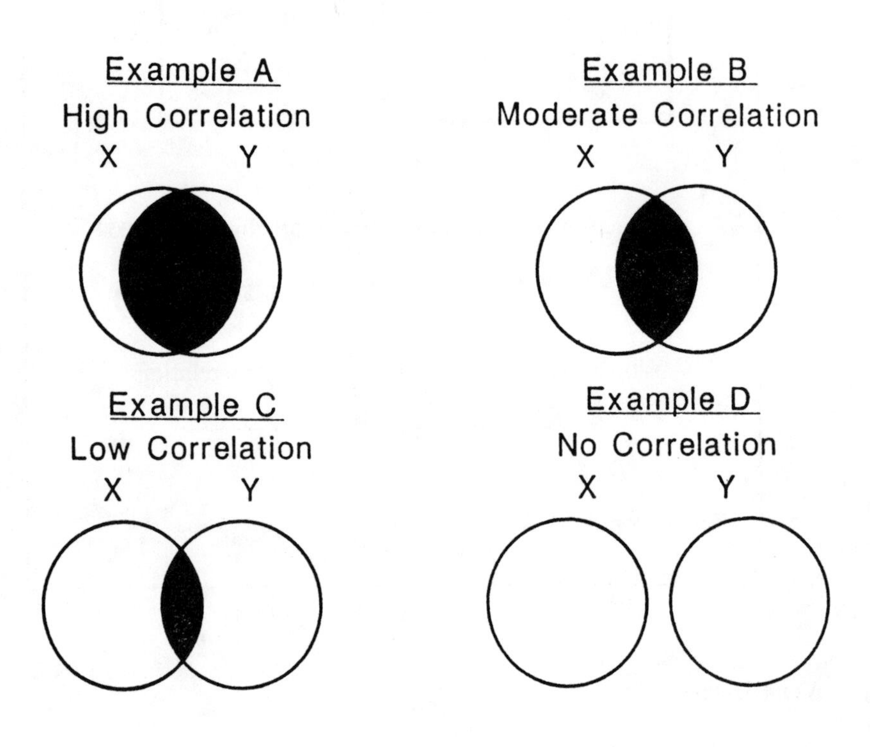


Figure 16. Venn Diagrams of different degrees of correlation.

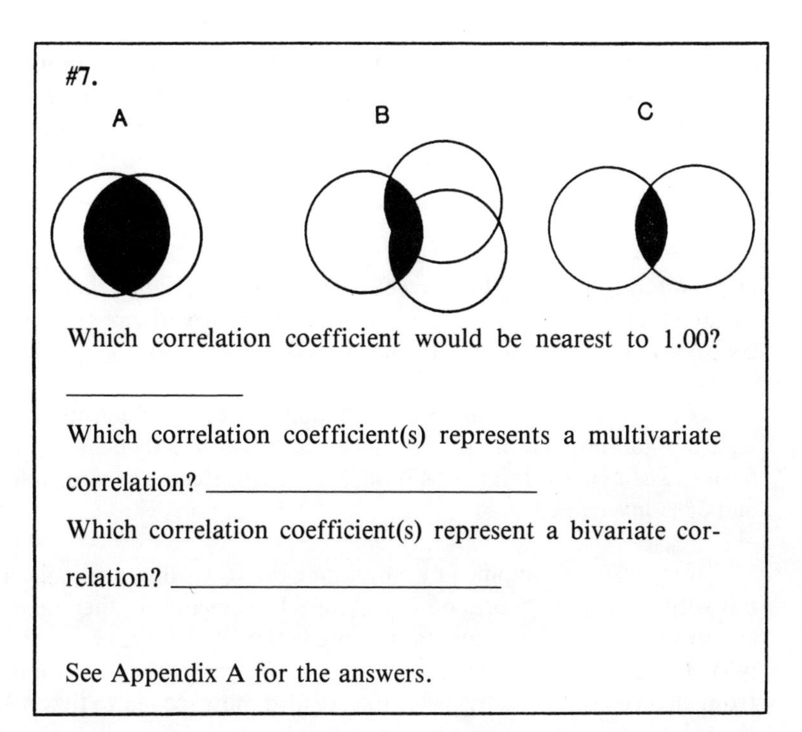


## INTERPRETATION

When the *correlation coefficient* is +1.00, -1.00 or 0.00 interpretation is clear and unambiguous. However, correlations of 1.00 rarely occur when applications are made in the "real world" of measurement. When dealing with *coefficients* of *correlation* smaller than 1.00 (e.g., r = .80) the most common misinterpretation is to interpret the coefficient as a percent (i.e., r = .80 is misinterpreted as an 80% relationship). This misinterpretation is probably an understandable attempt by intelligent persons to transfer past mathematical knowledge to a new context. To interpret a *coefficient of correlation* (r) as a percent the *coefficient of correlation* must be squared ($r^2$) and then multiplied by 100. For example, an r of .80 when squared ($r^2$) equals .64 and then multiplied by 100 results in 64%. The quantity $r^2$ is called the *coefficient of determination*. But how do you interpret the *coefficient of determination* ($r^2$)? The *coefficient of determination* interpreted as a percent ($r^2$ x 100) can be thought of in several ways. Ways that will help clarify its meaning, help in interpreting the precision of various tests and help in understanding research about how well test scores predict other scores (e.g., how well a particular I.Q. test predicts reading achievement scores for a particular group of students.)

The 64% used in the preceding example ($r = .80$ and $r^2 = 64\%$) may be thought of as the percent of their total variance that the two variables have in common. The idea of overlap of variance (variance in common) of two variables is demonstrated in Figure 16.

If $r = .80$ was obtained by correlating the scores of a group on two separate administrations of the same test (Note 8), then in simplistic terms we can say the test has 64% accuracy (consistency) of measurement and 36% inaccuracy.

If $r = .80$ was obtained by correlating the scores on a particular I.Q. test with subsequent scores on a reading achievement test, then (again in simplistic terms) we can say that our prediction of these reading achievement test scores from the I.Q. test scores (for the group) is improved from chance prediction by 64%. But we also must recognize that a 36% chance of inaccuracy (error) of prediction exists.

If $r = .80$ was obtained by correlating the scores on a new I.Q. test with the scores on a well established I.Q. test (e.g., the Stanford Binet), then the author of the new test might claim a degree of validity (Note 9) (64% agreement with the criterion test and 36% disagreement).

**#8.**

The coefficient of determination for $r = .70$ is ________ ?

See Appendix A for the answer.

---

**Note 8:** In the following section on reliability we will learn to call this procedure *test-retest reliability.*

**Note 9:** In the following section on validity we will learn to call this a type of *criterion related validity.*

At this point you may be wondering why so much emphasis is being placed on converting *correlation coefficients* into *coefficients of determination.* A multiplicative function is involved in the computation and the interpretation of correlation coefficients. A common misinterpretation of a correlation coefficient is made by interpreting the numbers (values) as percents. If an r of .80 is interpreted as a percent, then logically it would seem that an r of .40 would be half as good as the r of .80. However the $r^2$ (coefficient of determination) for r of .80 is .64% and the $r^2$ for the r of .40 is only 16%. Easily one can see that .16% is not half of .64%. Use of the *coefficient of determination* makes clear the relative proportional value of different correlation coefficients.

The qualifier "general terms" was used in the definition section when describing high, moderate, and low correlations. When a correlation coefficient is used to describe the reliability (consistency) of a particular test, the kind of test being considered must be stated before a judgment can be made about whether that reliability coefficient is relatively high or low. For example, a reliability coefficient of .75 for a standardized I.Q. test is relatively low because these tests usually have reliability coefficients of .90 or higher. A reliability coefficient of .75 for a projective personality test is relatively high because projective personality tests often have reliability coefficients lower than .75. In the "real world" of tests and measurements, *coefficients of correlation* used to indicate how reliable a test is, how valid it is, or how well it predicts some other measure must be judged both generally and relative to the type of tests being considered.

A second common misinterpretation of a correlating coefficient is that it indicates a cause-and effect relationship. For example, if a correlation of .60 is found between the heights and weights of a group of adults, a person might jump to the conclusion that variations in height cause variations in weight. Actually differences in heredity and environment are generally recognized as causing variation in both height and weight. Because two different measures are correlated does not mean that one causes the other (e.g., intelligence and reading achievement or smoking cigarettes and lung cancer); the assumption of causality is not proven by a high correlation. Experimental research (not correlational research) is used to attempt to prove a cause-and-effect relationship. You may remember that the warning on the side of a package of cigarettes once stated that it had been determined that smoking cigarettes "may be" hazardous to one's health. This somewhat tentative warning was made on the basis of correlating the incidence of lung cancer, emphysema, and other respiratory illness with the number of cigarettes smoked per day. A later warning states that it "has been determined" that smoking cigarettes is hazardous to one's health. The later rather definite warning is based on the results of experimental research. Much of this research was performed using dogs as the experimental subjects. Dogs were placed in an aparatus that resembled the aparatus used in the old Pavlov studies. The difference was that the experimental dogs were restrained so that they had to breath large amount of cigarette smoke. The control dogs did not breath the cigarette smoke. As a result, significant numbers of experimental dogs developed cancer, emphysema, or respiratory illness. Humans were not used for this experimental research because of ethical considerations.

The terms *positive* and *negative* (and their signs + and -) tend to be misinterpreted when applied to *correlation coefficients.* The signs are used to indicate only the direction of the relationship (shown in Figure 14). If the relationship is a direct one (+), then high scores on one variable are associated with high scores on the other variable. If the relationship is an inverse one (-), then high scores on one variable are associated with low scores on the other variable. Positive does not necessarily mean "good" and negative does not necessarily mean "bad." Although a teacher might expect high I.Q. scores of students to be associated with high final test scores (positive relationship), he/she might also expect high I.Q. scores of students to be associated with low measures of cheating behavior (negative relationship). For purposes of prediction the larger the coefficient of correlation, regardless of sign, the better it is (e.g., $r = -.90$ is better for prediction than $r = .80$).

When using correlation coefficients for prediction, remember that the relative indicated accuracy of the prediction is for the entire group and not any individual within the group (except when $r = \pm 1.00$).

## KINDS OF CORRELATION COEFFICIENTS

Two types of bivariate correlation coefficients are generally used when dealing with test reliability and test validity and when using tests to predict other measures (e.g., school achievement). These two are the

*Pearson product-moment coefficient of correlation* and the *Spearman rank order coefficient of correlation* (Note 10).

Whenever a correlation coefficient is referred to without the type being specified you should assume that it is a *Pearson product-moment coefficient*. The *product-moment coefficient* is computed using the values (scores) as they appear in the distributions. It is generally assumed that these values have (or approximate) the characteristics of either an interval scale or a ratio scale (Note 11). The *rank order* coefficient is computed using ranks (i.e., first, second, third, etc.). If the author played a round of golf with three professional golfers (e.g., Tom Watson, Jack Nicklaus, and Andy Bean) he would undoubtedly finish fourth, (rank order). If the scores (total number of strokes taken) were compared, the measurement distance between the first three golfers would probably be quite small and the measurement distance between the third and the fourth (the author) ranked golfers would be quite large. This example demonstrates that ranks are less precise than actual scores. Therefore, the *product-moment* coefficient is generally preferred over the rank order coefficient. However, except for precision of the measurement scale, both types of coefficients are interpreted in the same way. The *rank order* coefficient is generally considered to be somewhat suspect (because of the less precise scale of measurement) and is interpreted more cautiously.

The common symbol for the *Pearson product-moment coefficient of correlation* is r. The *Spearman rank order coefficient of correlation* may be indicated by rho, by p (the Greek letter for rho), or by $r_{ranks}$ (r subscript ranks) depending on the inclination of the person doing the writing.

The *product-moment coefficient of correlation* may be called a bivariate linear regression (Note 12) technique when dealing with predicting one variable from another variable. Two or more variables (called

---

**Note 10:** Karl Pearson developed the *Pearson product-moment coefficient* and Charles Spearman developed the *Spearman rank order coefficient*.

**Note 11:** Nominal, ordinal, interval and ratio scales will be discussed in the following section *Assumptions Affecting Correlation*.

**Note 12:** Remember the straight line concept discussed earlier in the Definition section of correlation. The regression line is the line that best fits the relationship actually found between the two variables.

prediction or predictor variables) can be used to predict another variable (called the criterion variable). The correlation coefficient used for this kind of prediction may be called a *multiple correlation coefficient,* a *multivariate correlation coefficient,* or a *multiple regression correlation coefficient.* The Symbol commonly used is R (capital letter R). In Figure 17 is shown a Venn diagram of a *multi-variate correlation coefficient* using X and Y to indicate predictor variables and Z to indicate the criterion variable.

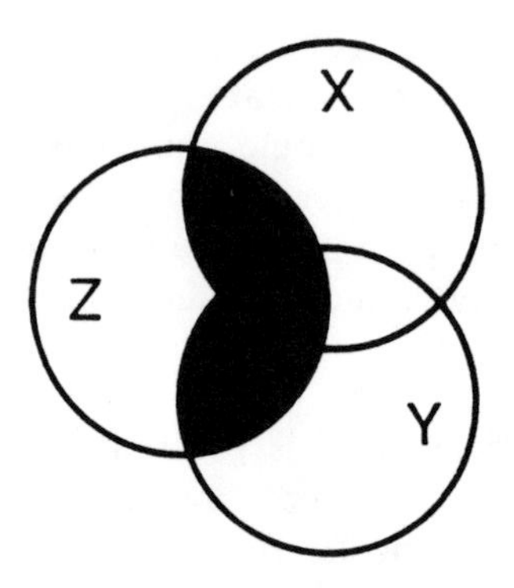


Figure 17. Venn Diagram of a multivariate correlation coefficient ($R_{zxy}$).

Use of this process of correlating several variables simultaneously may be clarified by the following example. Call variable X (in Figure 17) the Scholastic Aptitude Test (SAT) verbal scores, varible Y the SAT quantitative scores and variable Z the final college grade point averages. In combination the two prediction variables (SAT verbal scores and SAT quantitative scores) correlate higher with the criterion variable than either does individually. Thus $R_{z,xy}$ is larger (near to 1.00) than either $r_{zx}$ or $r_{zy}$. You may remember that several measures were required for your college entrance (e.g., SAT verbal, SAT quantitative, high school grade point average, high school rank, etc.). Regression analyses (both bivariate and multi-variate) computed for a particular group also yield regression weights that may be used together with scores for the prediction variables for subsequent individuals or groups to estimate how well they will probably achieve in a particular college (Note 13). Basically an R (e.g., R = .80) is interpreted like an r (e.g., r = .80) is interpreted. The essential difference is that the R represents correlating three or more variables simultaneously while r represents correlating only two variables.

Although multi-variate correlation is an extremely complex and abstract process this simplified discussion should provide some insight into how several scores may be used in combination to predict some criterion value. Today, the use of multiple predictors is very common in the "real world."

Several other bivariate and multivariate correlation techniques are used for special purposes. Resultant coefficients are interpreted in a similar fashion to the more common correlation techniques discussed previously. Since you may encounter some of these correlation techniques in your reading, the following brief and oversimplified explanations are provided. Again, this material provides only an introduction to statistics to help you become a "better informed" (but not completely informed) consumer.

Four of these special bivariate correlation techniques are the *biserial correlation coefficient,* the *point biserial correlation coefficient,* the *tetrachoric correlation coefficient,* and the *phi correlation coefficient.* These techniques are applied when one or both of the variables are not continuous variables, but are instead *dichotomous variables.* A *dichotomous variable* is one which is divided into only two categories. Two examples of such dichotomous variables would be sex (female or male) or credit versus no credit (where work usually graded A, B, or C should be designated as credit and work usually graded D or F would be designated as no credit). Sex is a naturally occurring or true dichotomous variable. Credit versus no credit is an artificially dichotomized variable. Table 6 shows the types of variables (continuous true dichotomous or artificially dichotomized) appropriate for each of these special bivariate correlation techniques.

---

**Note 13:** Your high school counselor may have used a publication from the Educational Testing Service (ETS) to predict how well you might do at a particular university based on your weighted SAT verbal score, SAT quantitative score, and your HS rank compared to past applicants to the particular college.

Table 6. Four Special Bivariate Correlation Techniques and Type of Variables

| Correlation Technique | Type of Variables | |
|---|---|---|
| | Variable X | Variable Y |
| Biserial Correlation Coefficient | Continuous | Artificially dichotomized |
| Point Biserial Correlation Coefficient | Continuous | True dichotomy |
| Tetrachoric Correlation Coefficient | Artificially dichotomized | Artificially dichotomized |
| Phi Correlation Coefficient | True dichotomy | True dichotomy |

A multivariate correlation technique that has often been used to study what certain tests (e.g., intelligence tests) measure and whether these tests measure more than one dimension (factor) is called *factor analysis.* For example a new intelligence test is divided into two major parts with each of these parts containing five sub-tests. Part one is labeled verbal ability and part two is labeled non-verbal ability. The rationale for the test is based on the assumption that each of these abilities is independent of the other. In other words each ability represent a separate *factor* of intelligence as measured by that particular test. To statistically investigate this assumption all possible intercorrelations of the ten sub-test scores of the two major parts are computed. If the five sub-test scores of part one are highly correlated with each other and if the five sub-test scores of part two are highly correlated with each other, and if the sub-test scores of part one and sub-test scores of part two are not correlated with each other to any significant degree, then this would be interpreted as evidence supporting the assumption that the test measures two separate dimensions or *factors*.

However, if all ten of the sub-test scores were found to be highly correlated with each other, it would be interpreted as evidence that the test measured only one dimension or *factor* of intelligence.

Of course some other pattern of highly correlated sub-test sets (which were independent of each other) might be found indicating three or more *factors*.

Obviously this extremely simple presentation of factor analysis only provides a general idea of one very complex and sophisticated statistical technique. *Path analysis* and *meta analysis* are two examples (out of many) of other sophisticated correlation techniques which require a great deal of statistical training to apply and interpret correctly and are best left out of this material for beginners.

## ASSUMPTIONS AFFECTING CORRELATION

Computational statistics courses and textbooks contain considerably more than computational formulas. Any statistical procedure (e.g., *product-moment correlation coefficient*) requires that data used meet certain assumptions in order to be appropriate. Even though most of these assumptions must be left for the time when (if?) you take courses in computational statistics, several assumptions need to be discussed briefly to help you better understand and critically evaluate various measures.

The *product-moment coefficient* is based on the assumption that the pattern of plots in the scatter diagram is best described by a straight line (e.g., see A & B in Figure 15). In some instances the pattern of plots in the scatter diagram may be best described by a curved line (*curvilinear* rather than linear). For example, the scatter diagram in Figure 18 show a curvilinear relationship between a measure of attitude toward English (the academic subject) and a measure of English ability.

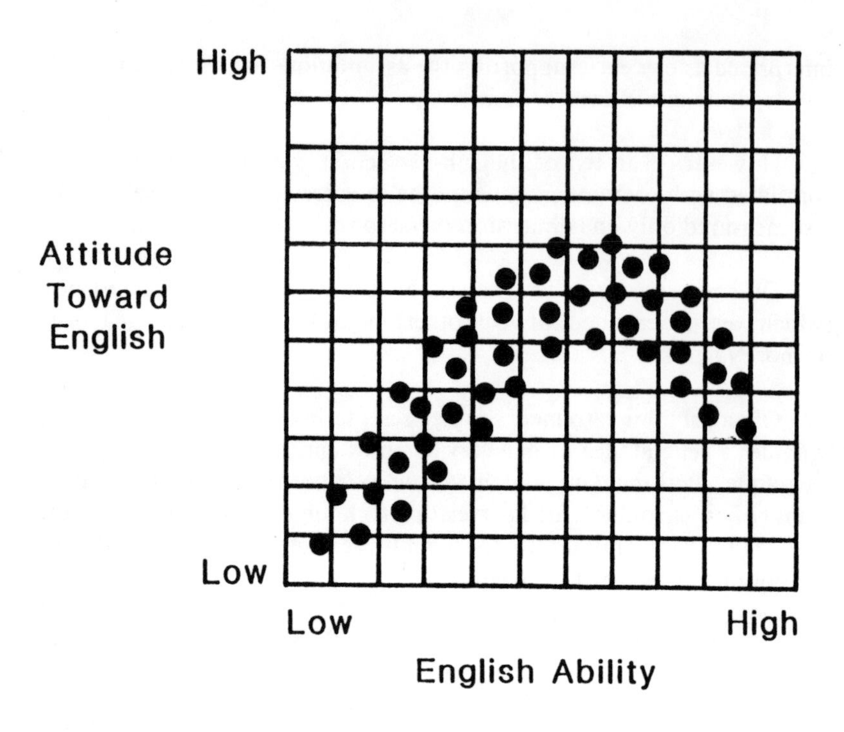


Figure 18. Scatter diagram for measures of attitude toward English and English ability.

The first part (low to moderate ability) of the scatter diagram indicates a fairly high positive correlation. But the second part (moderate to high ability) shows a fairly high negative correlation. Because of the effect of negative correlation diminishing the positive correlation (or vice versa) the computed r would be nearer 0.00 correlation than the true relationship between the two measures and thus not appropriate. Certain correlational techniques are appropriate for curvilinear relationships but they are not presented in this beginning level material.

The usual assumption is that the two measures to be correlated are independent (of each other) measures. For example, if you correlate the subsection scores of a test with the scores for the total test, the r will probably be too high (*spuriously high*). If you were to correlate the I.Q. scores for a group with those same scores, the r should be 1.00 (provided the computation is done correctly). Because the test subsection scores partially determine the total test score, the resultant r will be nearer 1.00 (spuriously higher) than it would have been if the two scores were independent measures.

Earlier you were told that the assumption was that the variables used to compute the product-moment coefficient were (or approximated) interval or ratio scales. The rank order coefficient was computed from ranks (ordinal scale). Earlier you learned to classify numerical evidence (measurements) as either *discrete* or *continuous.* In addition measurements may be classified (described) as *nominal, ordinal, interval,* or *ratio scales.*

*Nominal scales* may be described as pseudoquantitative (pseudomeasurements) because they represent the assignment of numbers to different unordered discrete categories. For example, you might assign the number 1 to the category female and the number 2 to the category male if you wished to code data for inclusion on a computer program. The numbers only name (nominal) the categories and have no quantitative meaning (i.e., number one *does not* mean that females rank first and number two *does not* mean that males are twice as good as females).

*Ordinal scales* generally indicate that the individual possessing the greatest measured amount of an attribute (e.g., weight, I.Q., etc.) is ranked 1, the second greatest amount is ranked 2 and so on. However, you may not assume that the difference between ranks 1 and 2 is necessarily the same as the difference between ranks 2 and 3. For example, imagine that you have 6 people line up at one end of a city block and place judges at a finish line at the other end of the block and then fire a blank pistol to start the race. You would expect that the contestants in the race would not all cross the finish line at the same time and probably the distance (interval) between the runners finishing first, second, third, fourth, fifth, and sixth would vary. If the age, weight, and physical condition of the six runners was quite different, then the intervals would undoubtedly be quite different. So ordinal scales can be rather crude measurement scales. The *rank order coefficient* which is computed from ranks is considered to be less precise than the *product-moment* coefficient which is computed from scales where each interval on the measurement scale is considered to be equal to every other interval on the measurement scale.

*Interval scales* have the attribute of each interval along the measurement scale being equal to every other interval. This added precision is found in test scores expressed as standard scores (e.g., deviation I.Q.s, SAT scores, etc.). However, because *interval scale* scores do not have an absolute zero it *is not* appropriate to say that a deviation I.Q. of 160 is twice as good as a deviation I.Q. of 80.

*Ratio scales* have the attributes of *interval scales* and also have an absolute zero. For example, height has an absolute zero plus even measurement intervals. Therefore, a person six feet tall may be described as twice as tall as the child who is only three feet tall. Undoubtedly few educational and psychological variables have ratio scale characteristics.

The number of cases (i.e., persons for whom we have scores on variables X and Y) is important in terms of the confidence (i.e., the probability that the r indicates a "real" relationship) you wish to place on a particular correlation coefficient. Generally, the larger the number of cases the greater the confidence you will place on the correlation coefficient. Applications of this assumption will be considered in the following sections of this material.

**#9.**

Which of the following correlation coefficients would be best for the purpose of prediction: **A.** r = .80; **B.** r =- .91 **C.** r = -1.13? ____________________.

You would expect the correlation coefficient for age of adults (subjects ages range from 40 years old to 80 years old) and speed of reaction time to be positive (+) or negative (-)?

____________________.

The correct correlation coefficient to compute for amount of money spent and amount of sales tax paid would be the Spearman rank order coefficient ($r_{ranks}$) or the Pearson product-moment coefficient (r)? ____________________.

The correlation between chronological age (subjects ages range from 2 years old to 70 years old) and reaction time would probably be positive (+), negative (-) or curvilinear?

____________________.

See Appendix A for the answers.

MODULE **IV**

# INFERENTIAL STATISTICS

Modern statistical inference is relatively new. Textbooks by R. H. Fisher (Statistical Methods for Research Workers, 1925, Oliver and Boyd, Ltd., Edinburgh, England) and George W. Snedecor (Calculation and Interpretation of Analysis of Variance and Covariance, 1934, Collegiate Press, Inc., Ames Iowa) stimulated applications of modern statistical inference in the agricultural and biological sciences. Applications to the behavioral sciences and to education have been more recent. Greater availability of high speed electronic computers has stimulated further development of modern statistical inference and has accelerated its application to many fields of endeavor.

The following material provides an introduction to some aspects of inferential statistics frequently used in the behavioral sciences. Fundamental concepts, that are logical extensions of the descriptive statistical concepts you have just learned, are featured. Understanding these fundamental concepts should aid your critical evaluation of validation studies of tests as well as other research and evaluation studies in the behavioral sciences.

## ESTIMATION

Inferential statistics is used for the purpose of making generalizations about a large group (*population*) from numerical evidence available from only a part (*sample*) of the large group. Means, standard deviations, correlations, and other values for a *sample* are called *statistics.* Means, standard deviations, correlations and other values for populations are called *parameters. Statistics* are used to estimate *parameters.* For example, if you wish to know the mean age of all graduate students attending Ball State University (BSU) during the 1985 first summer session, you could compute the mean age of a sample of those students. The mean age of the sample (*statistic*) would then provide the best available way of estimating the mean age of the population (*parameter*).

Gathering data and computing *statistics* for a sample is obviously more feasible (easier and less expensive) than gathering data and computing parameters. In some instances populations are not finite. For instance, the generalization that the distribution of human intelligence forms a normal curve involves an infinite population. An impossibility would be to measure the intellectual functioning of all persons who ever lived, who live now, and who will live in the future. Certainly intelligence tests cannot be administered to dead people or to people who have not yet been born. To administer the same I.Q. test to everyone who lives now also is not really feasible. The cost and time needed for such testing would be prohibitive and cultural and language differences would probably cause the results of any one I.Q. test to be less than valid. However, when an I.Q. test is administered to a large and somewhat representative sample the resultant distribution of scores tends to be approximately symmetrical. Therefore, the generalization is that, if the infinite population could be measured, any non-normal aspects of the distribution of scores *probably* would be averaged out, resulting in a normal distribution.

Obviously the more the sample is like (representative of) the population the greater the likelihood (*probability*) that the estimate is accurate. Sampling is an integral part of modern inferential statistics. Actually modern inferential statistics developed from the classical theory of sampling.

## HYPOTHESIS TESTING

Suppose that you observed that the male graduate students in the previous example (mean age of BSU graduate students) seem to be older than the female graduate students. Is this probably a function of the sampling or is it probably a true difference? This question can be stated as a *directional hypothesis* (a hypothesis is a tentative assumption stated as a generalization). That is, male graduate students at BSU are older than female graduate students at BSU.

For the inferential statistical process called hypothesis testing, the *directional hypothesis* is usually restated as a *null hypothesis*. The *null hypothesis* is not a negative hypothesis, but rather one postulating no difference between the groups. That is, there is no statistically significant difference in the ages of male and female graduate students at BSU. This approach is used so that probability inferential statistical techniques may be used to estimate the odds (probability or chance) that observed (measured) differences between the groups differ by a large enough margin from no differences between the groups to be a function of "true" differences rather than just chance differences. Another way of explaining the *null hypothesis* is to say that any observed differences in the ages of male and female graduate students at BSU is probably a function of sampling error (chance) and does not represent a "true" difference.

To test the *null hypothesis,* the means and variances (standard deviation squared) for the two groups (male and female) are compared. This type of inferential statistical test is called an *analysis of variance* statistical test. The acronym for *analysis of variance* is ANOVA. In Figure 19 is graphically represented the *null hypothesis* (A), an instance where no statistical significance difference is found (B), and instance where the observed difference is statistically significant (C).

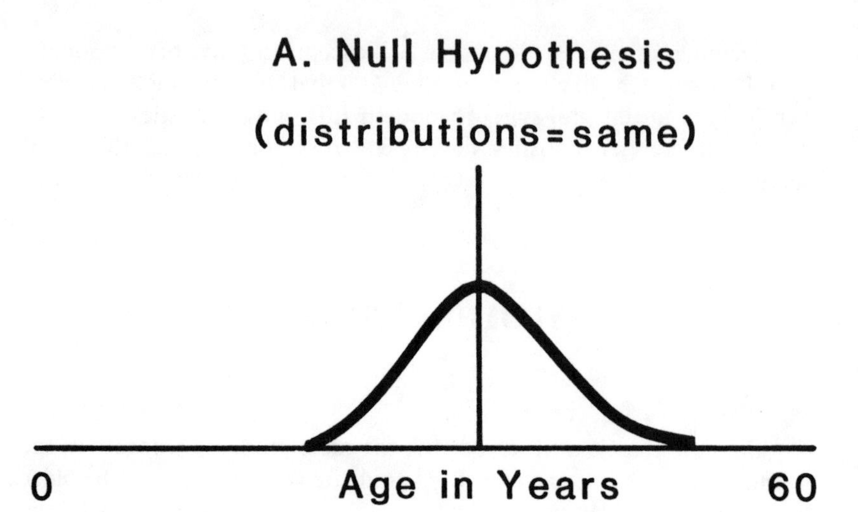


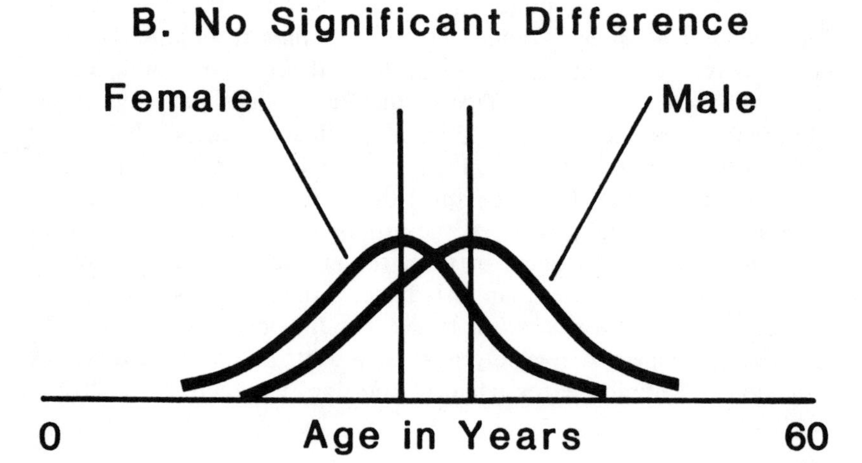


Figure 19. Test distributions.

Figure 19. Continued.

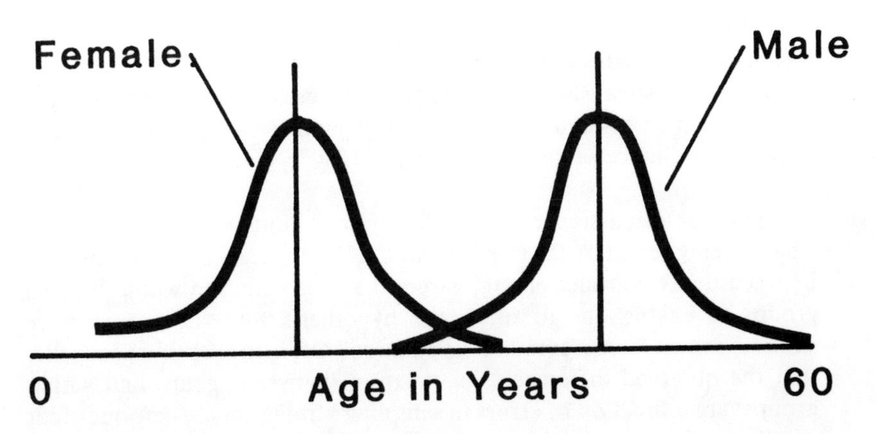


ANOVA tests involving two groups are called t-tests and ANOVA tests involving three or more groups are called F tests. The tests are also referred to as *t ratios* and *F ratios*. A t ratio is obtained by dividing the between groups variance by the within groups variance (Note 14).

$$t = \frac{\text{Between Groups Variance}}{\text{Within Groups Variance}}$$

The *between groups variance* is a function of how far apart the two means are. The farther apart the two means are the greater their variance around their common mean (the average of the two means).

The *within groups variance* represents the average variance within the two groups. Each group has a variance (standard deviation squared) and the mean (average) variance for the two groups is called the *within groups variance*.

In Example B (Figure 19) the between groups variance is relatively *small* and the *within groups variance* is relatively *large* resulting in a very

---

**Note 14:** Remember variance was earlier defined as the average of the squared deviation scores or as the standard deviation squared.

small t ratio (*nonsignificant*). In Example C (Figure 19) the *between groups variance* is relatively *large* and the *within groups variance* is relatively small resulting in a large t ratio (significant).

The greater the variation between groups as compared with the within group variation, the larger the t ratio and therefore the greater the chance of statistical significance. That is the greater the probability that the observed (measured) differences are true differences and not just a function of chance sampling differences.

Perhaps discussion of the following questions will tend to clarify what is being tested by the t test of statistical significance. Is the variation between the two groups enough larger than the variation within the two groups to enable one to reject the hypothesis that the groups were samples drawn from a common population? Or, is the more logical belief that the observed differences in variation (between groups and within groups) are a function of errors in sampling? In Figure 20 these questions are illustrated using dotted lines to represent common population (real or theoretical) and using solid lines to represent samples.

In Part A of Figure 20 the odds (chance) of drawing two homogeneous samples so divergent and in the tails of the population would be extremely large and the t test would be statistically significant. Therefore you reject the null hypothesis (no difference between sample variances) and conclude that it is more likely that each of the samples is from the center of a different population. This is logical because small homogeneous samples probably come from the center of a population distribution.

In Part B of Figure 20 the two samples are so similar that the t-test would not be statistically significant. Therefore you fail to reject the null hypothesis and conclude that probably the two samples are from a common population. Any differences are only chance sampling errors.

The two samples in Part B of Figure 20 are probably near the center of the common population because the odds strongly favor drawing two very similar homogeneous samples from the part of the population where the largest number of cases fall. The chance of drawing one homogeneous sample from one tail of a population is quite remote and odds against drawing two similar homogeneous samples from one tail of a population are astronomical.

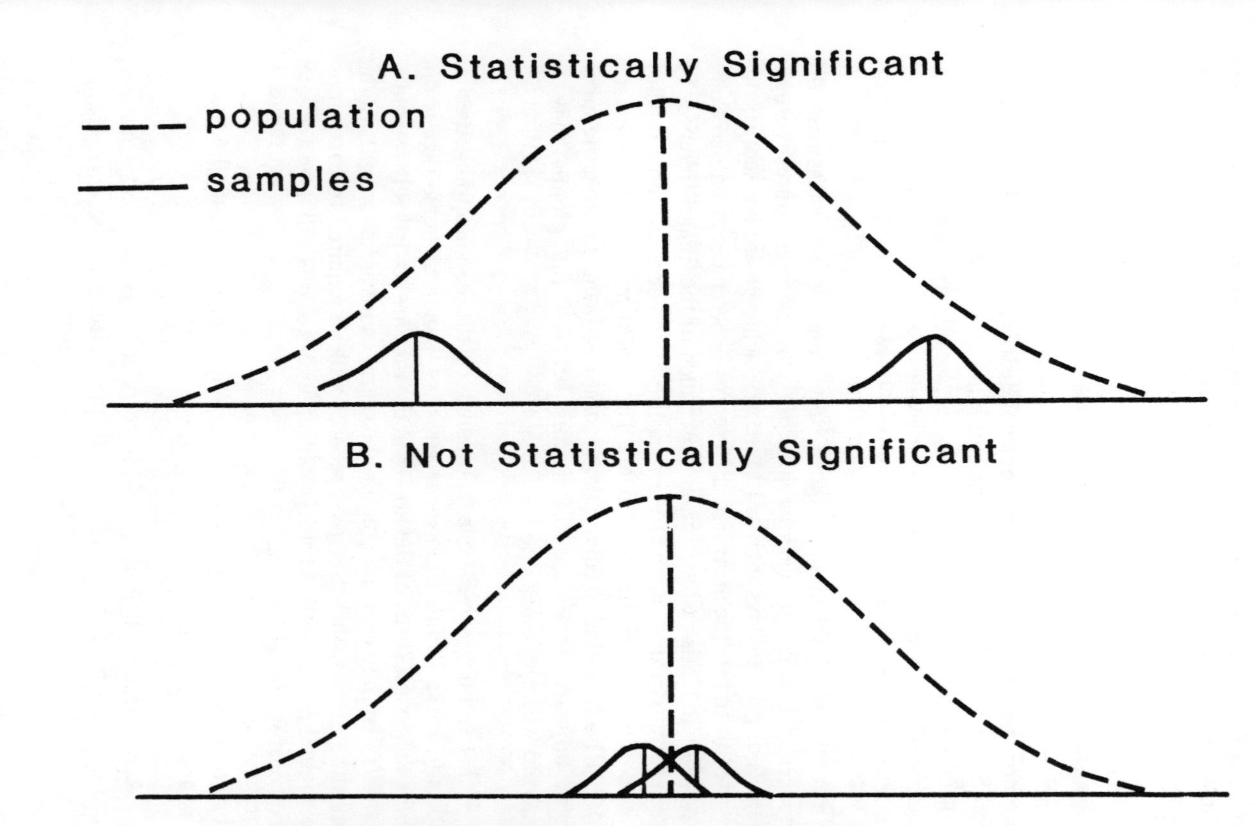


Figure 20. Relationship of samples to common population.

The t-test was used to explain ANOVA hypothesis testing because it is simpler to deal with two groups. However, the basic concept can be expanded for three or more groups (F test).

The F test ANOVA is used for research situations where three or more groups are employed. The assumptions of interval (or ratio) scale data, homogeneous variance, and normal distributions pertain. To insure homogeneous variance and normal distributions, random selection of subjects for the various groups and random assignment to treatments is usually employed. However, in some cases the researcher cannot use random selection and must use intact groups. When this situation occurs the researcher may use F test *analysis of covariance (ANCOVA).*

Analysis of covariance statistically controls for measured differences between the various groups. Covariate measures (usually pretests or premeasures such as I.Q. scores, achievement test scores, etc.) are used to equate the measured differences between the groups so that it is possible to statistically estimate how much of the final differences between groups is due to covariate measure differences and how much is due to the experimental treatment. The *multivariate correlation technique* (R) is first used to equate the covariate influences so that the subsequent adjusted treatment differences between the groups may be analyzed with the F test.

This technique has been compared to the handicapping used in sports to make competition more equal. A handicap horse race is one where the past records of the horses in the race are statistically treated so that different horses will be required to carry more or less weight. In this way the horses with better past records will be required to carry more weight and presumably the horses with poorer records will have a greater chance of winning the race.

The F test ANCOVA is interpreted in the same way that all ANOVA tests are interpreted.

Again testing hypothesis with correlation is different (probably more difficult to understand) than testing differences between groups. The null hypothesis for a correlation coefficient is stated differently. It may be stated that "the correlation coefficient (r) is 0.00," or that "there is no correlation between the measures being correlated". If an r of .40 is found to be significant (either by computing the significance or by consulting an available statistical table of the significance for r) the conclusion is that a true correlation exists. However, an improper conclusion would be that the true correlation coefficient for the population is necessarily r = .40. The computed r = .40 (statistic) is simply the best currently available estimate of the true r (parameter).

Statistical tests such as t-tests, F tests, r, and R are classified as *parametric* statistical tests. The term parametric is applied because these tests involve using *statistics* to make *inferences* concerning populations and their *parameters.* The appropriate use of these tests involves certain assumptions.

Some of the assumptions that may help you critically evaluate whether the correct statistical tests were used are as follows. The measurement data used was either interval or ratio data. For differences between groups the samples used had normal distributions and were relatively homogeneous. For correlation a linear relationship existed between the variables rather than a curvilinear relationship between the variables. If these assumptions are not met, the results of the statistical test used may be erroneous. For example, the true t ratio may actually be much larger or smaller than the t ratio computed.

Statistical tests employing *nominal* or *ordinal* measure data are called *non-parametric* statistical tests (Note 15). Since these tests do not technically involve generalizing from statistics (samples) to parameters (populations) they are labeled *non-parametric.* Therefore, the assumptions of normal distribution and homogeneous variance are not requirements for *non-parametric* tests of statistical significance.

The Spearman rank-order-correlated coefficient (rho or $r_{ranks}$) is a *non-parametric* correlational statistical test. The assumption of a linear relationship still applies, but the quality of the measurement data does not measure up to the assumed quality of interval or ratio scale data.

The Mann-Whitney U Test, which uses ordinal data, is a fairly common *non-parametric* test of the differences between groups. It has been called a *non-parametric* equivalent of the t-test. It is applied when the assumptions of normal distributions and/or homogeneous variance cannot be met.

---

**Note 15:** Note that interval or ratio data can be very easily be converted to ordinal data. However, the process of converting ordinal data to interval data is quite complicated and to convert ordinal data to ratio data may well be impossible.

Chi-square ($\chi^2$) compares observed (actual) frequencies in two or more groups with expected or hypothesized frequencies in these groups. For example, you might use chi-square to test the hypothesis that "there was no difference in the proportion of males and females who completed high school at Schlochville High School during the 1984-85 academic year." This study involves the use of frequencies (enumerations) of dichotomous data. The dichotomy concerned is completed high school or did not complete high school. In the chi-square analysis only the frequency of the completed high school part of the dichotomy will be used. When the 1984-85 academic year seniors entered high school as freshmen there were exactly 200 males and 200 females. Only 300 of these original 400 students are included on the graduation list. Three hundred have completed high school and 100 have not completed high school (assuming those who transferred have been accounted for). If no difference exists in the proportions of males and females completing high school, we would expect that 150 males and 150 females would be represented on the graduation list. However, if we found that there were 190 males and only 110 females included on the list, the magnitude of the chi-square test would be large enough for us to reject the null hypothesis and conclude that significant difference existed and that a significantly higher proportion of males (compared to females) completed high school in Schlochville during the 1984-85 academic year. Studies of drop-out rates for various groups, rate of achieving academic distinction for different groups, and similar studies have often used the chi-square statistical technique.

Generally non-parametric tests are not as powerful as parametric tests. In simple terms this indicates that a parametric test, generally, will reject a null hypothesis more often than a comparable (one designed to carry out the same function) non-parametric test.

If you have persisted this far, the idea that probability is generally an integral part of inferential statistics and specifically a key part of hypothesis testing must be developing in your mind. This idea is correct and use of such words as odds, chance, and probability were deliberately used to foster the idea. The "world" of inferential statistics is a relative "world." You should not expect absolute proof. The best that can be hoped for is obtaining what you consider to be acceptable odds. What are acceptable odds? Are there generally accepted levels of probability? These questions lead to the matter of defining *statistical significance* and considering whether statistical significance always corresponds with "real" (practical significance as applied in the "real world") significance.

## STATISTICAL SIGNIFICANCE

*Statistical significance* is one application of a larger classification called *confidence limits.* In addition to statistical significance, *confidence limits* deals with various standard errors (e.g., standard error of measurement, standard error of estimate, etc.) (Note 16).

*Statistical significance* reflects the odds that observed differences (or a correlation coefficient) represents a "true" difference (or correlation) rather than a chance difference. How confident you must be (what odds are acceptable) in order to conclude that differences are "true" differences is a value judgment. Traditionally in the behavioral sciences the .05 level of significance has become the minimum acceptable level to conclude "true" differences and reject the null hypothesis. The .05 level of *statistical significance* indicates that the odds are 5 (or fewer) chances in 100 that observed differences could be attributed to chance. Another way of expressing this idea is to say that the odds are 95 chances in 100 that the observed differences are "true" differences. Tables of statistical significance of t, F, and r usually show the .05 and the .01 levels of statistical significance and some also include the .001 level of significance. The .01 level of significance indicates one chance in 100 that observed differences are chance differences and the .001 level of significance indicates one chance in 1000 that differences can be attributed to chance. These levels of statistical significance were arbitrarily set and by continued use over time have become the traditionally acceptable levels.

Suppose members of your class were invited to attend a special presentation by a noted researcher, Dr. Vivian LaRue-Calabash (Note 17), a medical doctor. She describes a series of experimental studies testing the hypothesis that a newly discovered vaccine will prevent the development of certain kinds of cancer. The research studies she describes have all been conducted using laboratory animals, (e.g., rats and certain members of the monkey family). After the formal presentation and question and answer period, Dr. LaRue-Calabash explains that

---

**Note 16:** Standard error of measurement will be included in a later section on reliability.

**Note 17:** To the author's knowledge no person of this name exists, the name was invented for this example.

the next step in the research process will be a series of experiments using human subjects. As a result of the animal research studies she has received all the necessary governmental and institutional permission clearances for the series of experimental research studies with human subjects. If she stated that the probability (odds) of the vaccine preventing cancer rather than producing cancer are about 95 in 100 (.05 level) would you be willing to volunteer? Even if the probability level (odds) was 999 in 1000 most people would not be eager to volunteer. Indeed with these odds (probability levels) permission for the use of human subjects probably would not have been given.

Areas such as medical research set more rigorous levels of statistical significance. When life or death is concerned, 5 chances in 100 of being in error would not be acceptable to most people.

## TYPE I AND TYPE II ERRORS

In testing the statistical significance of observed differences (or correlation coefficients) the possibility exists of being in error when rejecting (or failing to reject) the null hypothesis. *Type I error* occurs when the null hypothesis was rejected when it was true (no true difference exist). A *Type II error* occurs when the null hypothesis is accepted (failure to reject) when it is false (true differences really exist).

**#10.**

Setting the level of statistical significance at the .05 level (or the .01 or the .001 level) rather than at a higher level (e.g., .10 or .20 level) guards against making a type ______ error.

See Appendix A for the Answer.

*Statistical significance* and *"real" significance* may not always be the same thing. *Statistical significance* deals with the odds that findings are not the result of chance, while *real significance* refers to how useful findings are when applied in the "real" world. For example, a correlation coefficient (r) between I.Q. scores and reading achievement scores of .195 is *statistically significant* if computed for the scores of a group of 100 students. The coefficient of determination for an r of .195 would be .038. It is obvious that an I.Q. test explaining less than 4 percent of the variance in reading achievement scores would not be a particularly practically useful predictive instrument in this instance ("real" significance). The time and expense of administering, scoring, etc. such a test would be judged to be too high in relation to the usefulness of the resultant test scores.

**Degrees of Freedom(df)**

An in depth treatment of *degrees of freedom* assumes a fairly high level of mathematical sophistication. Therefore, complete discussion of *degrees of freedom,* including the rationale for determining them, is not included in this material. Computational statistical courses (and textbooks) usually contain a more complete coverage of determining *degrees of freedom* for each statistical test of significance.

This discussion of *degrees of freedom* is aimed at helping you read tables of statistical significance. In order to achieve this objective some general definition information and some information concerning *degrees of freedom* for *ANOVA,* chi-square, and r are provided.

Mathematical definitions of *degrees of freedom* deal with the number of elements (within a mathematical process) that can vary and still permit arriving at a specific outcome. For example, an addition problem requires that the sum of four (4) numbers result in a total of twenty-seven (27). The value of three (3) of the numbers may vary freely because the fourth number can be the necessary value to make sure that the sum of the four (4) numbers is twenty-seven (27). In this case the *degrees of freedom* are N - 1 (4 - 1) or three (3). Three (3) of the numbers can vary freely but one (1) cannot.

Application of the mathematical meaning of *degrees of freedom* to inferential statistics (sampling, estimation, and prediction) is designed to insure more accurate outcomes. For example, the estimate of the probability of making a Type I or Type II error when testing an hypothesis is more accurate than it would have been if the *degrees of freedom* had been ignored.

*Degrees of freedom* for ANOVA tests of significance indicate the number of groups and the size of the groups involved in the test of significance. For example an $F_{2,60\ df}$ indicates that three groups were involved. The 2 (of 2,60df) is determined by subtracting one from the number of groups (degrees of freedom for groups is the number of groups minus one). The 60 (of 2,60df) refers to the size of the groups (number of individuals involved). Because the rationale for determining these degrees of freedom for the size of groups is quite involved we will leave that for later study of computational statistics. However, you should remember that as sample sizes increase so does our confidence in the resultant F (Note 18). If you examine an ANOVA table of statistical significance, you will see that as the *degrees of freedom* for size of the groups increases the size of the F or t necessary for significance at a particular level of significance (e.g., .05 level of significance) decreases. The higher the level of statistical significance (i.e., the .01 level as compared to the .05 level) the larger the required F or t value.

Often the degrees of freedom for the number of groups are not included for t-tests (i.e., $t_{40f}$). A t-test always involves only two groups, so $t_{1,\ 40df}$ is shortened to $t_{40df}$ with the 1 for the number of groups being assumed (understood).

**#11.**

Which of the following would be the appropriate F ratio for a research study with three experimental groups and one control group?

A. $F_{1,80df}$

B. $F_{2,80df}$

C. $F_{3,80df}$

D. $F_{4,80df}$

See Appendix A for the answer.

---

**Note 18:** You may remember from the earlier presented material that large sample size results in a diminished effect of unusual scores due to the averaging process.

*Degrees of freedom* for chi-square ($\chi^2$) indicates the complexity of the chi-square table. One degree of freedom indicates the simplest table and two or more *degrees of freedom* indicates a more complex table. The formula for chi-square *degrees of freedom* is computed by multiplying the number of rows (in the $\chi^2$ table) minus one, times the number of columns minus one (df = rows minus one times columns minus one). So a table comparing males and females (2 columns minus 1 = 1) on expected and actual proportion of dropouts (2 columns minus 1 = 1) would result in one degree of freedom (1 for rows times 1 for columns, i.e., 1 times 1 = 1). The simpler the chi-square table (smaller number of degrees of freedom) the smaller the $\chi^2$value necessary for statistical significance at a particular level of significance (e.g., .05 level of significance). The more complex the chi-square table (larger number of degrees of freedom) becomes the larger the $\chi^2$ value necessary for statistical significance at a particular level.

The number of pairs of scores used in computing r serves the function of degrees of freedom. The larger the number of pairs of scores the smaller the r necessary for statistical significance at a particular level of significance. For example, with 10 pairs (N = 10 or in some instances the degrees of freedom are stated as N - 1 or 9 in this instance) the r must be at least .632 to be statistically significant at the .05 level of significance, while an r of .312 computed for 40 pairs of scores (N = 40) is statistically significant at the .05 level of significance.

**#12.**

Which of the following F ratios would be most likely to be statistically significant?

A. $F_{2,20df}$

B. $F_{2,30df}$

C. $F_{2,40df}$

D. $F_{2,50df}$

See Appendix A for the answer.

# COMMENCEMENT NOTE

Congratulations you have worked your way through this conceptual approach to basic concepts of descriptive and inferential statistics. Because of the complexity and abstractness of these ideas, you should periodically restudy the material. Redundancy is a necessity when dealing with material that is both complex and abstract.

You may have already discovered that resource instructional materials and/or resource people (instructors or colleagues who understand this applied brand of statistics) can help you to expand your understanding of this material. If not, you should now consider talking with resource people and consider studying some of the beginning level conventional statistics textbooks.

The following sections are provided to demonstrate how statistics is applied to measurement and evaluation (tests and measurements) and to research. Measurement is used in schools, business, the armed forces, etc., and is a basic part of empirical research. The quantification of observed differences (measurement of variables) from individual to individual, from place to place, and from one time to another time has many practical uses. Statistics helps to organize, describe, and draw inferences about the observed differences or similarities (correlation).

The following sections apply what you have learned to the evaluation and interpretation of the kinds of measurement instruments and scores used in research, evaluation, and day-to-day decision making in the behavioral sciences.

MODULE **V**

# RELIABILITY AND VALIDITY OF MEASUREMENT INSTRUMENTS

Unfortunately most of the measurement used by behavioral scientists is not as precise as measurement of height and weight (*ratio scales*). Measurement of intellectual functioning (I.Q. tests), learning, personality, etc. results in only *ordinal* or *interval* scale data, not *ratio* scale data.

Intelligence, learning, personality, and motivation are called *hypothetical constructs*. Hypothetical constructs are abstractions and do not exist in concrete form. We do not know for sure that these things actually exist. We have used logic to attempt to describe them and hypothesize (make tentative guesses) about how they are structured (organized) and that they do, indeed, exist. As a result of their abstract (not concrete) nature these hypothetical constructs must be measured indirectly. For example, achievement (a measure of change in achievement test scores over an interval of time, e.g., over the time spent of one academic year of schooling) is measured in order to estimate the degree to which learning has taken place. Learning is inferred from achievement test results.

Achievement tests usually contain only a sample of all of the possible questions that could be asked. As a result failing to answer any of the questions correctly does not necessarily indicate zero (no) achievement. The assumption that any one question in the test represents the same difficulty and discrimination (measurement difference) levels as every other item in the test is not necessarily true. Therefore the raw scores on such a test can only be considered to be *ordinal* scale data. For example, the author sometimes notices that two students sitting beside each other have exactly the same score on a class (achievement) test. Curiosity, rather than a suspicious nature, motivates the author to examine, item by item, the two test papers. Was it just a coincidence that the two scores are the same or was some other variable, such as peripheral vision, partially responsible. Quite often items passed or missed on the two papers are very different even though the total raw scores are the same. Obviously these scores would place both people in the same rank order in relation to the entire class but for them to represent exactly the same level of precise measurement is extremely doubtful.

In order to convert the *ordinal* data to *interval* data the raw scores must be expressed in terms of their standard deviation from the mean score of the group. The process of expressing raw scores in terms of standard deviation units results in *standard scores* (Note 19).

The quality of the measurements obtained using any measurement device is important. The idea of your butcher using a scale that measures inaccurately, especially if it indicates that a steak weighs more than it actually does, is not attractive considering the high price of meat per pound. The effect of indirect measurements expressed as *ordinal* or *interval* data is to highten the concern about the quality of those measurements.

---

**Note 19:** Remember that earlier you learned that standard deviations represent standard units along the measurement continuum. *Standard scores* will be covered in the following section on interpreting scores.

The major aspects of the quality of measurement instruments are validity and reliability.

## VALIDITY

*A measurement instrument is valid if it measures what it purports (claims) to measure.* This general definition needs to be qualified in several ways. First, validity is always in relationship to who is being tested, where, and when. If a test was validated on a national sample of white subjects during 1950, its validity when administered to a group of Sioux Indians in South Dakota during 1980 is questionable.

Second, validity of a test is dependent upon that test having reasonably good reliability. A test must be reasonably reliable or it cannot be valid. However, test reliability does not guarantee validity (Note 20).

A validation study of a test is a study carried out to determine the relative validity and reliability of the test. Correlation is the major statistical technique used in validation studies.

Essentially four types of test validity are used. These types are *construct validity, concurrent validity, predictive validity,* and *content validity.* Each type of test validity represents a particular emphasis or approach to demonstrating the validity of a measurement instrument.

The example of a hypothetical new I.Q. test, the Miller Quick Scoring Intelligence Test, should assist in understanding validity. This new test consists of a tape measure, a felt tip pen, and an electric hair clipper. The test is administered by carrying out the following sequential steps.

First, the felt tip pen is used to make an X on the forehead of the subject directly above the bridge of the nose and parallel to the eyebrows. This X is the first reference mark. A second reference mark will be made at the back of the head. This mark will be made on the small bump in the skull directly above the hollow running vertically along the back of the neck. However, before this mark can be made, the electric hair clippers

---

**Note 20:** Reliability of tests will be discussed in detail following the material concerning test validity. For now, reliability refers to the consistency of repeated measurements.

must be used to clip away the hair around the head parallel to the two reference marks. Following the hair clipping and the marking of the second reference mark (X) the tape measure is used. The tape measure is carefully placed around the head passing over the two reference marks. The number of inches (and fractions of an inch) constitutes the score on this I.Q. test.

The relative validity (Note 21) of this hypothetical I.Q. test depends on the answer to two questions. Will this instrument measure in a reasonably reliable (consistent) fashion? Does this test really measure intellectual functioning (intelligence)? Because intelligence is a hypothetical construct the construct validity of this test is a relevant concern.

**Construct Validity**

*Construct validity* refers to the degree to which the structure of the test corresponds to (matches) the characteristics (structure) of the particular hypothetical construct which the test (measurement instrument) purportedly measures. Often a logical and largely nonstatistical comparison (nonstatistical correlation) of the various characteristics of the test with the characteristics of the hypothetical construct is presented. If the characteristics of the test and the characteristics of the construct correspond to a large degree, then the logically assumption is that the test has a reasonably high degree of construct validity. This deductive approach is often used for instruments such as intelligence tests and personality assessment measures (Note 22).

To demonstrate the kind of evidence and logic often used when construct validity of a measurement instrument is being considered, consider the hypothetical Miller Quick Scoring Intelligence Test (MQSIT).

The fact that the MQSIT scores tend to result in a normal distribution curve when this test is administered to a large and rather representative (typical) group of people seems to be consistent with the

---

**Note 21:** Validity really is not an absolute (valid vs. not valid) term but a relative term (how relatively valid).

**Note 22:** Advanced statistical correlational techniques (e.g., factor analysis) are also used to explore the degree of construct validity of tests such as the WISC-R (Wechsler Intelligence Scale for Children), etc. A special kind of factor analysis (confirmatory factor analysis) is currently being used to study the construct validity of tests (e.g., intelligence tests).

generalization that intelligence is normally distributed in the infinite population. Of course this is not conclusive evidence, because measures of things other than intelligence may also tend to form a normal distribution curve.

The idea that the measurement of the outside circumference of the head is an indirect measure of the circumference of the brain (one of the possible measures of size of the brain) might be linked to the findings of certain research studies. The reports of positive correlations between brain size and I.Q. and the reports of periodic increases in brain size corresponding to the developmental sequence and timing of theoretical formulations for cognitive development (e.g., Jean Piaget's theoretical formulation) could be interpreted as evidence of construct validity. However the research studies investigating the correlation between brain size and intellectual functioning defined brain size in terms of weight (usually the measures of brain weight was made at an autopsy) rather than circumference. A brain of greater mass (weight) can be contained in the same sized skull cavity, as one weighing less, so the difference in operational definition of brain size makes a great deal of difference in the attempt to link these research findings to the construct validity of the MQSIT.

This limited example shows the nature of the attempt at building a logical case for the correspondence of the structure of the test and the structure of the phenomenon (intelligence in this instance) under consideration. The evidence presented together with the logic used often fails to convince some people that a particular instrument is indeed valid. If the measurement instrument purportedly measures a phenomenon by applying a technique which is part of a particular theory, then the acceptability of that particular theory affects the willingness of some people to accept the validity of that instrument. For example, a projective personality test (e.g., an ink blot test) is linked with Freudian theory. When a person is asked to express what a relatively ambiguous stimulus (e.g., on ink blot) reminds them of, the notion of psychological projection becomes an issue. A person must agree with the notion of psychological projection (one of the dynamisms or defense mechanisms postulated in Freudian theory) in order to accept the construct validity of a measurement instrument using such a technique. Because some people do not accept Freudian theoretical formulations, they probably do not accept the validity of projective personality tests.

Because the *construct validity* approach to establish the validity of a measurement instrument may be somewhat unconvincing to some people, then other approaches may be more convincing (either alone or in combination with construct validity). But, what kind of evidence tends to be convincing to most people?

Many educated people are impressed by the results of empirical research findings. A fairly common developmental sequence of how people come to value research findings as the key to understanding the nature of various phenomena is as follows.

As a small child a person may decide that one of his/her parents is the prime source for any desired bit of knowledge. That parent usually is replaced as the "all knowing" authority by another person (often a teacher) at some later point. For example the author's son thought (for a while) that his father was the key source for information outside the home. One fall morning when father and son went outside to get the car in order to go to school and to work, a heavy frost covered the ground and the car. The well meaning, but somewhat pedantic father, attempted to explain the difference between frost and snow. The son was apparently unimpressed with the distinction between the white stuff forming on a surface rather than falling on the surface. He remarked that the frost scraped from the windshield of the car felt like snow. Despite a repeat mini-lecture on the key difference, the son remarked as they reached his school that he would ask the teacher whether or not it had snowed. The tenure of the author as all knowing authority was obviously almost over.

Eventually the veracity of what people say comes into question. That which is printed, particularly that which is printed in a textbook, is judged to be true. But conflicting accounts of topics like the American Revolution compared to an account of the insurrection in the colonies results in a loss of faith in the objectivity of different authorities.

What does the research literature indicate? This question indicates a certain faith in the results of empirical research findings. Of course, a person usually discovers that a distinction exists between good research and research that is not so good. But empirical research studies to attempt to establish the validity of a measurement instrument tend to be convincing to many people. *Concurrent validity* and *predictive validity* involve using empirical research studies to establish the validity of a measurement instrument.

### Criterion Related Validity

*Concurrent validity* and *predictive validity* may be grouped under the heading of *criterion related validity*. The statistical procedure used for *criterion related validity* is correlating the results of the test you are attempting to validate with the results of some *criterion measure.*

If the author wished to empirically establish the validity of his hypothetical new I.Q. test, the Miller Quick Scoring Intelligence Test (MQSIT), he might administer it to a large sample of adults and then administer the Wechsler Adult Intelligence Scale (WAIS) to the same sample of people. In this instance the MQSIT is the measure to be validated and the WAIS is the *criterion measure.* Remember the MQSIT is the facetious hypothetical I.Q. test consisting of a tape measure, a felt tipped pen, and an electric hair clipper. The felt tipped pen is used to place an X between the eyebrows and a second X on the bump directly above the hollow at the back of the neck. Of course hair presents a biasing factor (a factor causing error of measurement) and the hair clipper is needed to clip off the hair around the head parallel with the two X marks. When the hair is removed the tape measure is placed parallel around the head over the two X marks and a measurement (score) is attained. The larger the circumference of the head the higher the implied I.Q. This measure of hat size (with the hat pulled down so that the wind would not blow the hat off the head) really does not measure I.Q. but in the past equally strange measures of I.Q. have been attempted. However, if those people with large heads achieved high I.Q. scores on the WAIS and those with average size heads received average WAIS I.Q.s and those with small heads attained low WAIS I.Q.s, a positive correlation would result. This empirical evidence would then be interpreted as *criterion related* evidence supporting the validity of the MQSIT. Of course a low positive or negative correlation coefficient would indicate that the validity of the MQSIT instrument was in doubt. In this example the criterion related validity approach would be labeled *concurrent validity.* Because both measures (the MQSIT and the WAIS) occur in the same time (temporal) span.

If the MQSIT was administered to a group of incoming freshmen students and four years later the graduating cumulative grade point averages for these same students were gathered, then these two scores (MQSIT and cumulative grade point average) for the sample of students could be correlated. If so, resultant correlation coefficient would be called a *predictive validity* coefficient. The term *predictive validity* indicates that scores on one measure (an aptitude test) predicts scores on the criterion measure at some future time. In this example, if those persons with big heads earned 4.0 cumulative grade point averages and those with average sized heads earned 2.0 cumulative grade point averages and those with tiny heads flunked out of school, a high positive correlation coefficient would be interpreted as evidence supporting the *predictive validity* of the MQSIT. However, a small, negative, or zero correlation would indicate a lack of *predictive validity*. Since many authorities contend that intelligence tests (I.Q. tests) are primarily used to predict academic (scholastic) achievement, then the *predictive validity* of these tests is of prime importance.

**#13.**

What is most wrong with the following statement? "This test is valid."

A. The statement does not specify for what the test is valid.

B. The word "valid" is vague. A numerical coefficient should be given.

C. The statement is meaningless, since it does not specify the conditions of administration.

See Appendix A for the answer.

## Content Validity

*Content validity* refers to the degree to which the content of the text represents an adequate sample of the content universe assumed to be tested. For example, does a particular fourth grade arithmetic achievement test represent an adequate sample of questions from the various content areas, (e.g., addition, subtraction, multiplication, etc.) as well as an appropriate range of cognitive complexity (i.e., factual recall, problem solving, etc.)? A table of specifications (sort of blueprint) for an achievement test is generally prepared before the individual test items (questions) are written. The total number of questions that may be included in a particular test is limited by various constraints such as the age and ability level of those who will take the test. Generally a grid type *Table of Specifications* (see Table 7) is prepared in order to attain the desired balance (sampling) of tests according to both content areas and cognitive complexity.

Table 7. Table of Specifications for a Social Studies Test

| Teaching Objectives | Number of Items | | | |
|---|---|---|---|---|
| | Content Areas | | | |
| | 1 | 2 | 3 | Total |
| 1. Recall or recognition of information | 10 | 5 | 5 | 20 |
| 2. Interpretation of facts and generalizations | 5 | 10 | 5 | 20 |
| 3. Application of knowledge | 5 | 5 | 10 | 20 |
| Total | 20 | 20 | 20 | 60 |

Content Areas of the Civil War:

1. Historical aspects

2. Economic aspects

3. Political aspects

If you have an existing achievement test (e.g., a commercial standardized achievement test), you would use a Table of Specification in a somewhat different way. You would classify each item in the existing test according to content area and cognitive complexity and talley the number of items into the appropriate grids on a blank table of specifications. The resultant table would show the sampling of the available test

---

*Should be number of points if different numbers of points are given to some questions.

which could then be compared to the curricular emphases in your particular location (e.g., Schlockville, Indiana Public Schools). This process is called *content analysis.* In addition other analyses, such as reading level analysis of the questions, may be carried out to determine whether the reading difficulty of the questions in the test is appropriate for the students to be tested. This descriptive statistical approach provides a rather detailed analysis of the degree to which the existing test has *content* (curricular) *validity* for the students of the Schlockville, Indiana Public Schools.

*Content validity* is of prime importance for any type of achievement test, but also is important for instruments used in survey research (e.g., questionnaires and opinionnaires). In the case of survey instruments the content to be sampled is usually gotten from a review of the relevant literature or from a panel of experts in the area under consideration.

## RELIABILITY

If you were to measure something twice using the same measurement instrument would you get precisely the same measurement each time? If you administer the Miller Quick Scoring Intelligence Test to a group of subjects today and then administered the MQSIT to the same group tomorrow would the correlation coefficient for the two administrations be a perfect positive ($r = +1.00$) one? If you did get exactly the same measurement each time, you could say that the measurement instrument (in this instance) had *perfect reliability*. However, measurement instruments used by behavioral scientists are never perfectly reliable. Measurement instruments for measuring intelligence, personality, etc. do not measure as reliably (consistently) as do instruments for measuring height and weight. The observation has been that certain types of measurement instruments are generally more or less reliable than instruments of another type. For example, instruments to measure intelligence (I.Q. tests) are generally more reliable instruments than those used to measure personality.

Several ways exist to estimate the relative reliability of a measurement instrument. Each way has certain limitations. Some kind of statistical correlational process is used for each approach resulting in a *coefficient* (correlation coefficient) *of reliability*.

### Test-Retest Reliability

*Test-retest reliability* is determined by administering the test to a sample of persons and then readministering the same test to the sample at a later time. The scores on the first administration of the test are correlated with the scores on the second administration of the test. The resultant reliability correlation coefficient (r) is often called a *coefficient of stability,* because it shows how stable the scores remained over a span of time. Most definitions of test reliability are framed in terms of test-retest reliability. An obvious limitation of this approach is that time must pass before the retesting occurs and thus a span of time is needed to determine the relative reliability of the test. A less obvious limitation of the *test-retest reliability* approach is concerned with the meaningfulness of the questions in the test. If the questions are quite meaningful (i.e., they have a familiar structure and can be quite easily remembered), the the first administration of the test may have a different effect on how the various persons answer the questions on the second administration of the test, which could result in a change in the scores of some individuals on the second administration of the test. One way to possibly minimize this effect is to lengthen the time between the initial testing and the retesting.

### Alternate-Forms Reliability

*Alternate-forms reliability* (sometimes called parallel or equivalent forms reliability) requires having two or more forms of the test (e.g., Form A and Form B). To determine the relative reliability of the test Form A is administered to a group of persons and then Form B is administered to the same group. The scores for Form A are correlated with the scores for Form B resulting in a *reliability coefficient* (r) *of equivalence.* Having to construct a second test (second form) that covers the same content and measures exactly the same way as the first test, but used different questions represents quite an impressive limitation. For example, before 1960 the Stanford Binet Intelligence Test (a very carefully developed and standardized test) had two alternate forms (Form L and Form M). These two forms never correlated perfectly and in 1960 the revised single form test (the LM Form) was presented. This revised test used the best functioning items from the two previous forms (Note 23).

---

**Note 23.** You may wish to see the section about item analysis in a tests and measurement (measurement and evaluation) text book for information about such technical aspects of test items as item discrimination, item difficulty, and item validity.

## Internal Consistency Reliability

*Internal consistency reliability* refers to the concept of how consistent one half of a test is when compared with the other half of the test. It is assumed that it is possible to divide a test into equal (parallel) forms. Usually the division process is done by including the even numbered items in one half and the odd numbered items in the other half. The scores (for example, of persons) for the even half are correlated with the scores for the odd half resulting in a *split-half reliability coefficient* (r). Generally the greater the number of items in a test the higher (nearer 1.00) the reliability coefficient. Since the split-half reliability coefficient represents an estimate of the reliability of a test only half as long as the original test, it is an underestimate of the reliability coefficient for the original test. In order to estimate the reliability of the original test the Spearman-Brown "prophecy" formula is used. This formula is as follows:

$$\text{reliability of the original test} = \frac{\text{2 times reliability of half test}}{\text{1 plus reliability of half test}}$$

This approach requires only one test but involves considerable clerical work and necessitates computation of the Spearman-Brown estimate of the true reliability of the instrument.

Other approaches exist to examine the *internal consistency of a test.* Kuder Richardson Formula 20 (K-R 20) and Kuder-Richardson Formula 21 (K-R 21) are two statistical formulas to estimate the average (mean) of all possible split-half coefficients for a test. The K-R 21 assumes that all items in the test are of equal difficulty technical quality which makes the computation somewhat simpler than the computation for K-R 20. The K-R 21 formula tends to be a lower (more conservative) estimate of the test reliability than the K-R 20 estimate. These approaches together with other similar statistical approaches (e.g., coefficient Alpha) involve time-consuming data treatments but have become more common as computers have become more accessible.

All *internal consistency* approaches to estimating test reliability are inappropriate for estimating the reliability of speed tests (Note 24), but are appropriate for power tests (Note 25). To gain some insight into this limitation consider the following simple example. Suppose you administered a test with one hundred (100) different questions ranging from easy items at the beginning of the test to very difficult items at the end of the test (Note 26). One individual completes only the first thirty (30) items, a second person completes sixty (60) items and a third person completes ninety (90) items. Obviously the quickness (or lack of quickness) of these three persons has resulted in what amounts to quite different tests for each, both in terms of difficulty and length.

**Summary**

Reliability as well as validity needs to be qualified. The reliability of a measurement instrument is related to what (or whom) is being measured. Suppose you were asked whether a 60 inch tape measure or a 60 inch steel rule was a more reliable measurement instrument. The correct answer would be that it depends on what is being measured. The tape measure would yield more reliable (consistent) measurements on a curved surface and the rule would yield more reliable measurements on a flat surface. Therefore, the reliability coefficient for a particular test administered to the students in Schlockville, Indiana probably will not be of the same magnitude as the reliability coefficient for the norming sample of the test (as reported in the technical manual of the test).

---

**Note 24:** A *speed test* uses rather strict time limits to increase the discrimination between individuals taking the test. Clerical aptitude tests and many groups I.Q. tests are examples of tests which use speed as one aspect of gaining measurement discrimination.

**Note 25:** A *power test* provides generous time limits so that generally all persons have sufficient time to consider all questions.

**Note 26:** Achievement test, scholastic aptitude tests, etc. are usually constructed using a *gradient of item difficulty*. That is, easy items at the beginning of the test progressing through more difficult items to the most difficult items at the end of the test.

**#14.**

A. Match the terms in 1 through 3 with those in a through d:

| | |
|---|---|
| ______ a. Test-retest reliability | 1. coefficient of internal consistency |
| ______ b. Alternate-forms reliability | 2. coefficient of stability |
| ______ c. Split-half reliability | 3. coefficient of equivalence |
| ______ d. K-R Formula 20 | |

B. If the reliability of a reading test is r =.64, and if the length of the test is doubled, the reliability coefficient would probably ______________________ .

See Appendix A for the answers.

## STANDARD ERROR OF MEASUREMENT (Sem)

*Reliability* (precision) of a measurement instrument is reported as a correlation coefficient (reliability coefficient). As you learned earlier, correlation coefficients are difficult to interpret. This is particularly true when you wish to interpret an *obtained score* (Note 27) in terms of a range of scores within which the *true score* (Note 28) probably falls. The

---

**Note 27:** An *obtained score* is the score you get (obtain) when you administer the test to a particular person.

**Note 28:** A *true score* is the score you should get if the measurement instrument was completely reliable (reliability coefficient = 1.00).

*standard error of measurement* (Sem) is the statistical procedure that allows you to estimate such a range of scores. This information is particularly helpful when you are using test scores to aid in making classification (Note 29) (placement) or selection (Note 30) decisions.

In over simplified terms, the *standard error of measurement* (Sem) represents an estimate of the standard deviation of test scores a person would obtain if the test was taken a large number of times. The *standard deviation* of the test and *reliability coefficient* of the test are used to calculate (estimate) the *standard error of measurement* for the test. The formula for *standard error of measurement* is as follows:

$$\text{Standard error of measurement} = \text{Standard deviation of the test} \sqrt{\underset{\text{(perfect reliability)}}{1.00} - \underset{\text{(reliability of the test used)}}{r}}$$

If you had a test with a standard deviation of 15 points and a reliability coefficient of .75 the *standard error of measurement* would be 7.5.

$$\text{Sem} = 15\sqrt{1.00 - .75} = 15\sqrt{.25} = 15 \times .5 = 7.5$$

For a test with a standard deviation of 15 points and a reliability coefficient of .91 the *standard error of measurement* would be 4.5.

$$\text{Sem} = 15\sqrt{1.00 - .91} = 15\sqrt{.09} = 15 \times .3 = 4.5$$

Obviously as the reliability coefficient approaches 1.00 the standard error of measurement decreases. If a test was perfectly reliable (r = 1.00) the *standard error of measurement* would be 0.

Because the *standard error of measurement* is one type of standard deviation the percents under different portions of the normal curve are used for interpretation (see Figure 21). However, the percents for standard errors are probabilities (chances in 100) rather than percent of cases (frequencies) as they were for standard deviations of scores.

---

**Note 29:** *Classification* (placement) implies accepting all persons but assigning them to different classes, sections, etc.

**Note 30:** *Selection* implies that some persons are accepted and some are rejected. A selective admission program, such as a doctoral program in school psychology, would be an example of *selection*.

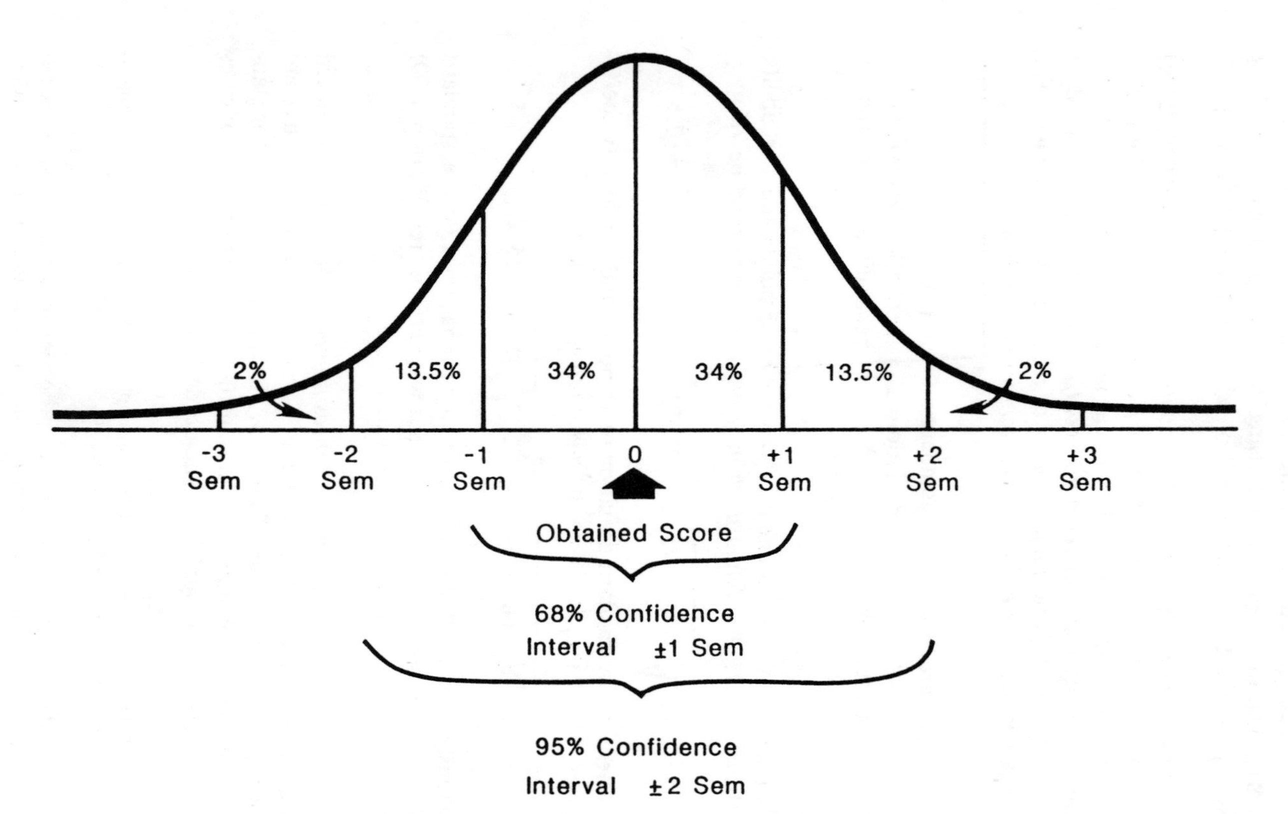


Figure 21. Confidence intervals for standard error of measurement.

The confidence interval of ± 1 Sem (the range between the score that fall one *standard error of measurement* above the obtained score and the score that falls one *standard error of measure* below the obtained score) is called the 68% confidence interval. The probability is 68 chances out of 100 that a persons true score falls somewhat between a +1 Sem and a -1 Sem from his/her obtained score. This is often expressed as 2 chances out of 3 that a persons true score falls between a ±1 Sem (Note 31).

The confidence interval of ±2 Sem is called the 95% confidence interval (Note 32). The probability is 95 chances out of 100 that a person's true score falls somewhere between a +2 Sem and a -2 Sem from his/her obtained score. Suppose a person's I.Q. of 72 was obtained by administering a test that had a Sem of 4 points. We would be 68% confident that that person's true score fell some place within the range from 68 I.Q. to 76 I.Q. We would be 95% confident that the persons true score fell some place within the range from 64 I.Q. to 80 I.Q.

The *standard error of a mean* (Se mean) and the *standard error of estimate* (Se est) are two other kinds of standard error frequently encountered when a person reads empirical research studies. Like the standard error of measurement both are closely related conceptually to the standard deviation.

**Standard Error of a Mean (Se Mean)**

The *standard error of a mean* is used to estimate how relatively representative, in terms of a particular variable, a sample selected is of the population from which it was drawn. For example, if we randomly select a sample of fifty (50) cases, will the mean I.Q. for the sample be an

---

**Note 31:** In the following section dealing with interpreting scores, the 68% conficence interval (± 1 Sem) is used to generate percentile band scores.

**Note 32:** The 95% confidence interval is actually ± 1.96 Sem but a bit of license is usually taken with precision in order to add ease of computation and interpretation.

accurate estimate of the mean I.Q. for the population? The *standard error of a mean* is estimated as follows:

$$\text{Standard error of a mean} = \frac{s \text{ (standard deviation of the sample)}}{\sqrt{N - 1} \text{ (number of cases in the sample)}}$$

Obvious as the sample size (N) increases and the standard deviation of the sample (s) decreases the standard error of a mean decreases. The smaller the *standard error of a mean* for a sample the greater our confidence in the proposition that the mean of the sample is an accurate estimate of the mean of the population and the sample is relatively representative of the population (at least in terms of the particular variable measured). You can easily understand why statistics from large samples are apt to be more representative than statistics for small samples.

**Standard Error of Estimate (Se est)**

The *standard error of estimate* ties the basic concepts about standard errors to correlation and the idea of the precision (lack of error) of using one variable (e.g., height) to predict another variable (e.g., weight) (Note 33).

You will remember that a *standard deviation* describes the deviation (variation) of scores around the mean of a distribution and a *standard error of measurement* or a *standard error of a mean* also describe deviations from an obtained score or from a mean. The *standard error of estimate* describes variability of plotted points (on a scatter diagram) from the straight line of best fit (straight line reporting a perfect correlation). If all the plotted points fall on the straight line of best fit, there is no error of prediction. However, if some of the plotted points deviate from the straight line of best fit (maybe referred to as the regression line), then the deviations represent error of prediction. The more the plotted points deviate (on the average) from the straight line of best fit the lower (nearer 0.00) the correlation coefficient and the larger the standard error of estimate and the less accurate (reliable) the predicted measures.

---

**Note 33:** A reminder that accuracy of prediction using correlational data should be viewed as prediction for a group and not prediction on an individual basis.

**#15.**

You are to choose between three group intelligence tests (Test A, Test B, and Test C) and the technical manual of each test indicates the following Sems for each test:

Test A Sem = 4.5

Test B Sem = 3.7

Test C Sem = 6.4.

You should choose test_________________________ .

See Appendix A for the answer.

MODULE **VI**

# INTERPRETING SCORES

Modern textbooks for courses in tests and measurements tend to contain the words *measurement* and *evaluation* in their titles. *Measurement* refers to the act of quantifying or determining differences in magnitude. We measure height, weight, academic achievement, and so forth. *Evaluation* refers to making value judgments about the scores obtained when we measure. For example when a medical doctor measures the height and weight of a particular patient, the doctor may conclude that the patient is grossly overweight for his/her height.

Generally when we measure (obtain a score) we compare the resultant score to either a *norm* or a *standard* in order to *evaluate* (make a judgment about how relatively good or bad the score is) that score. A *norm* is a measure of central tendency (usually the mean) for a comparison group (norm group). For example the doctor's judgment that the patient was grossly overweight resulted from the realization that the patient's weight was 2 standard deviations above the mean weight of the average person of the patient's particular height. This information was obtained by examining a growth table for adults (a type of norm table). This type of measurement is referred to as *norm referenced testing.*

If a person goes to a motor vehicle licensing branch to take a written driver's test, he/she may be told that a minimum passing score is 35 items answered correctly. The minimum score of 35 items answered correctly is a *standard.* This type of testing is called *criterion referenced testing.* How the particular *standard* (criterion level) was determined is not always clear. It may have been a purely arbitrary decision or it may have been generated from normative data. That is, it may be the number of questions answered correctly by 80% of those who have taken the test in the past. Other terms often associated with *criterion referenced testing* are *minimum competency testing* and *mastery testing.* In any case, a *standard* represents a predetermined standard (level) of performance.

**#16.**

If you wished to determine if a second grade student was ready to progress to a higher level (curricular content including vocabulary, etc.) reading textbook you would use a

______________________________ referenced test.

See Appendix A for the answer.

## STANDARD SCORES

Because *norms* are based on the performance of a group of people (norm group), the normal curve is used for interpretation. The assumption is that the norm group will be a representative sample of the population one wished to generalize about and that the instrument (test) used discriminates the relative amount (magnitude) of the particular attribute under consideration from person to person. If you study (or have studied) construction of norm referenced tests, you will realize that discrimination power (measurement power) is a prime concern. Fifty percent difficulty is the "ideal" difficulty level (how many persons answer correctly and how many answer incorrectly) for test items in a norm referenced test because this level of difficulty maximizes item discrimination.

Earlier you learned that standard deviations resulted in even intervals along the base continuum (abscissa) and partition values (e.g., percentiles) resulted in uneven intervals along the base continuum. Therefore, the normal curve and the percent of cases contained between standard deviations are used for the interpretation of *standard scores.*

The word *standard* in standard scores refers to standard deviation and not to standardized tests (Note 34). A *standard score* has a set mean and standard deviation. In Figure 22 are shown the more common standard scores and how they correspond to different points on the base continuum of the normal curve.

**Standard Deviation (σ)**

*Standard deviation* (σ) scores have 0 as the mean and most of the cases fall between a -3σ and a +3σ. The *z scores* are essentially the same except the σ sign has been dropped. If you are considering scores from a test with a mean of 100 and a standard deviation of 15 points, a score of 130 may be expressed as +2σ or a +2.O z. To calculate you subtract 100 (the mean) from 130 and have a remainder of +30 (+ because the score of 130 is larger than 100), which you then divide by 15 (the standard deviation). Because 15 goes into 30 two times, the score of 130 is a +2σ or if you prefer it may be expressed as a +2 z. If you add all of the percentages for the cases under the portions of the normal curve up to a +2σ (or +2 z), you will discover that nearly 98% of the cases fall at or below that point. Therefore, the 98th percentile rank (equivalent) corresponds to the test score of 130.

---

**Note 34:** *Standardized tests* have fixed directions for administration and scoring and have usually been constructed by professional test constructors. Although many *standardized* tests are norm referenced some also are criterion referenced.

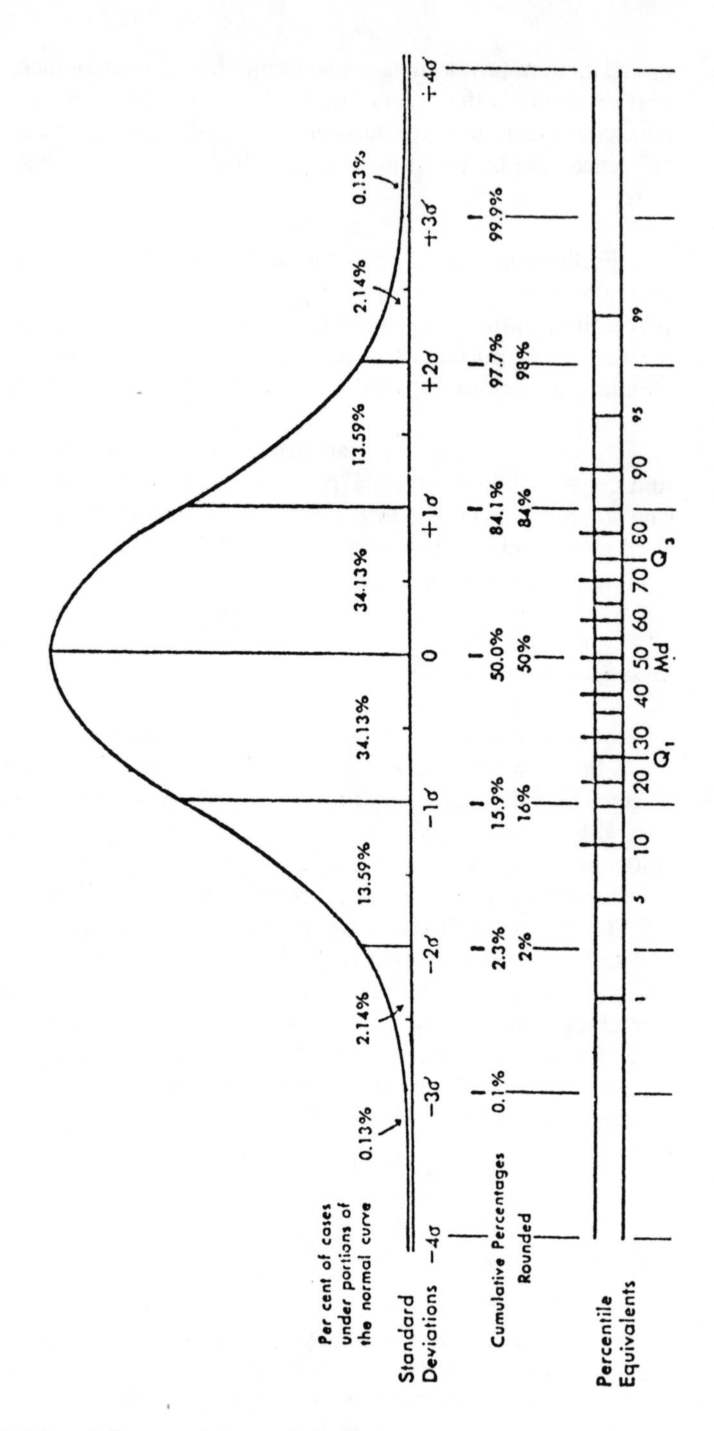

Per cent of cases under portions of the normal curve
0.13%
2.14%
13.59%
34.13%
34.13%
13.59%
2.14%
0.13%
Standard Deviations
-4σ
-3σ
-2σ
-1σ
0
+1σ
+2σ
+3σ
+4σ
Cumulative Percentages
0.1%
2.3%
15.9%
50.0%
84.1%
97.7%
99.9%
Rounded
2%
16%
50%
84%
98%
Percentile Equivalents
1
5
10
20
30
40
50
60
70
80
90
95
99
Q1
Md
Q3

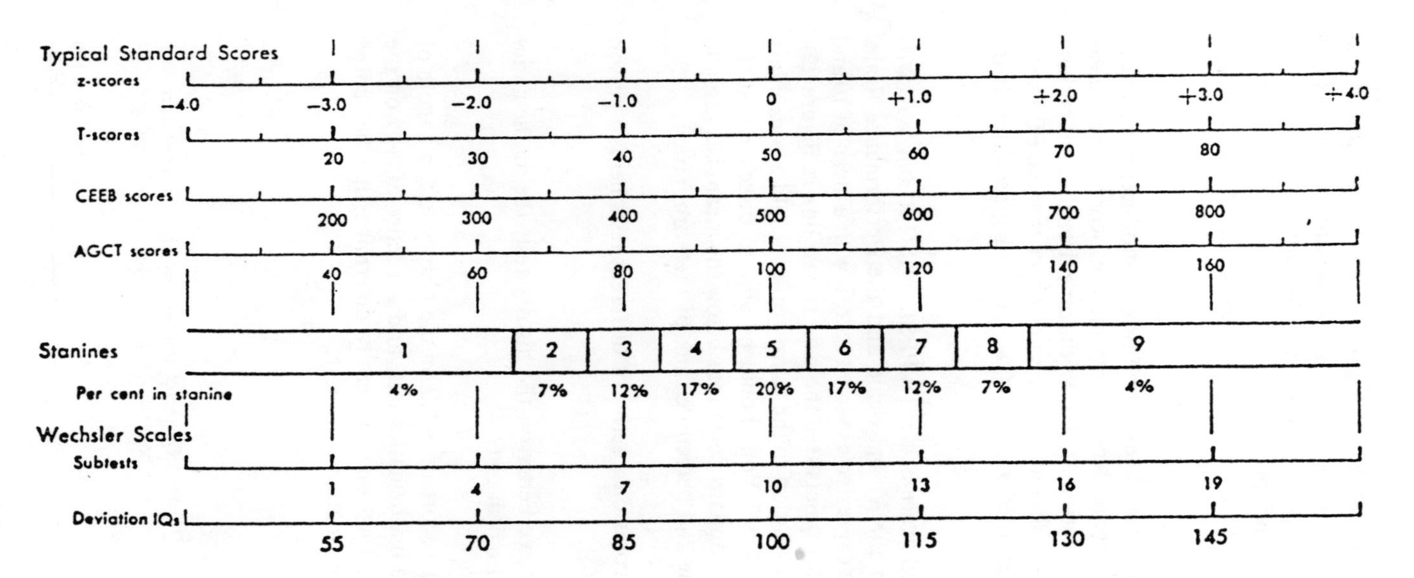


Figure 22. Standard scores and percentile ranks for the normal curve (The Psychological Corporation Test Service Bulletin, No. 48, 1955).

If you look at the bottom line of Figure 22, you will see that *deviation I.Q.s* (Note 35) have a mean of 100 and a standard deviation of 15 points. So the computation in the preceding paragraph demonstrates that a deviation I.Q. of 130 may be expressed as two standard deviations above the mean, as a z score of +2 or as approximately the 98th %ile of the norm sample.

CEEB (College Entrance Examination Board, producer of such tests as the SATs and the Graduate Record Examination) scores have a mean of 500 and a standard deviation of 100. With some thought (and perhaps a bit of computation) you should be able to discover that a CEEB score of 700 also may be represented as a $+2\sigma$, a +2 z and approximately the 98th percentile rank.

The Wechsler Intelligence Scales (WPPSI, WISC-R and WAIS-R) have subtests for both the Verbal Scale and the Performance Scale. These subtest scores are expressed as scaled scores, with a mean of 10 and a standard deviation of 3 points (see the next to last line in Figure 22). How relatively weak (in relation to the norm group) would a vocabulary subtest scaled score of 4 be? You should be able to discover that the scaled score of 4 is a -2 standard deviations below the mean, is a -2 z and is just slightly above the 2nd percentile rank of the norm group.

Having standard scores on a series of different tests helps a person answer questions such as:

1. Did Joe achieve better on the history test, the mathematics test, or the English test?

   If Joe had a z score of +1.0 on the history test, a z score of +2.0 on the mathematics test, and a z score of +.5 on the English test, then one can easily determine that he scored highest on the mathematics test.

---

**Note 35:** *Deviation I.Q.s* are I.Q. scores that are standard scores rather than *ratio I.Q.s* which are the ratio between mental age and chronological age.

$$\text{Ratio I.Q.} = \frac{\text{MA (mental age in months) times 100}}{\text{CA (chronological age in months)}}$$

2. Was Mary's achievement on the reading achievement test commensurate with her I.Q. score? Is she underachieving or overachieving?

   If Mary's I.Q. score was 130 (on the WISC-R I.Q. test) and her reading achievement score was 300 (on a reading test with a mean score of 300 and a standard deviation of 100), then her I.Q. score represents a z score of +2.0 and her reading achievement score represents a z score of 0.0. Therefore, Marys scores are not commensurate and reading achievement (as measured by the reading test) is considerably lower (two standard deviations lower) than her I.Q. Based on these data Mary has a sizeable underachievement.

**Stanines**

The label *stanines* is a contraction of the term standard nines. Each of the nine units represents one half of a standard deviation (Note 36). The midpoint (mean) of the normal curve falls exactly in the middle of the 5th stanine (or middle 20% of the cases). Any score which falls within the range for a particular stanine is expressed as the stanine score. For example, the deviation I.Q. of 130 could be expressed as being within the 9th stanine. *The major advantage of stanine scores is that they represent ranges rather than specific points* and thus may help to contend with the tendency of most people to view test scores as being more precise than they actually are. Remember that I.Q., achievement, and other similar tests are not perfectly reliable and therefore contain measurement error.

**#17.** A deviation I.Q. (the mean score = 100 and the standard deviation = 15) of 115 would correspond to a z score of ____________, a T-score of ____________, and would fall within the ____________ th. stanine and %ile rank of approximately ____________.

See Appendix A for the answers.

---

**Note 36:** Except stanine 1 and 9 which are obviously open-ended.

## PERCENTILE NORMS AND PERCENTILE RANKS

Another way to deal with the same concern is to express scores in terms of a *percentile band* (range) within the obtained score probably falls. *Percentile bands* are usually computed by dropping one *standard error of measurement* (-1 Sem) below the obtained score and going *one standard error of measurement* (+1 Sem) above the obtained score and reporting the percentile rank for each of these points as the lower and upper limits of the percentile band. If the *standard error of measurement* is relative (Note 37) large, then the percentile band will be rather large. Remember that a ± 1 Sem represents the 68% confindence interval. One is 68% confident that the true score (the score if there were no measurement errors) falls between a + 1 Sem and a -1 Sem.

A *percentile rank* is a point along the base continuum where raising a verticle partition (line or wall) would demonstrate that a certain percent of cases (individuals in the norm sample) falls at or below that partition. Thus a deviation I.Q. of 100 could be expressed as the 50th percentile rank. The scores of 50% of the norm sample fall at or below that point. People not sophisticated in measurement might misinterpret a percentile rank as the percent of questions answered correctly rather than a relative position compared to the norm group. Again, the uneven units along the base continuum (large at the extremes and smaller near the center of the distribution) of percentiles leads to difficulty when interpreting differences and/or changes in percentile ranks.

**#18.**

A standard percentile band (± 1 Sem) indicates that the odds (probability) are ____________________ out of 100 that the true score falls somewhere within the *percentile band.*

See Appendix A for the answer.

**Note 37:** When the Sem is quite large, it is logical to assume that the reliability of the instrument is relatively low and the precise interpretation of a score for that test is tenuous.

## OTHER CAUTIONS

Scores may be expressed in numerous other ways. Hopefully, the basic material you have just studied will enable you to work out correct interpretations of such scores.

For example, scores on standardized achievement tests are often reported in terms of *grade equivalent* scores. *Grade equivalent* scores are usually based on the assumption of a 10 month school year. So a *grade equivalent* score of 5.0 indicates the beginning of the fifth grade school year and a *grade equivalent* score of 5.5 indicates half way through the fifth grade school year. The *grade equivalent scores* represent a score relative to a norm group. So the *grade equivalent* score of 5.0 indicates the mean (average) level of fifth graders at the beginning of the school year. Assuming a normal distribution, you would expect that 50% of the students might score above grade equivalent and 50% might score below grade equivalent. Probably the most serious misinterpretation of *grade equivalent scores* is that they represent a *standard* rather than a *norm*. Then a person incorrectly assumes that all students should be at *grade equivalent level* or higher.

Scores on I.Q. tests may be expressed as mental age scores. Mental age scores are expressed in terms of the number of years and months of credit the individual has obtained when he/she took the I.Q. test. For example, a mental age of 6-2 indicates that the person answered correctly enough of the test items to attain a mental age score of 6 years and 2 months. This score indicates that the persons performance on the test was similar to the average performance of someone 6 years and 2 months old. If the person in question has a chronological age of 10 years 7 months then the best interpretation is that the person's performance is extremely low in relation to that person's chronological age. The assumption that this person will react precisely like the average 6 year 2 month old person is unrealistic. A better interpretation would be that the person's performance level is several standard deviations below typical performance by an average person of 10 years 7 months of chronological age.

Carefully examine the technical manuals of any test you use in order to determine the nature of the norming groups(s). The evidence concerning reliability, validity, standard error(s) of measurement, kinds of norms, and so forth is important for correct interpretation of the test scores.

You should now have a basic set of concepts that will allow you to better cope with the statistical aspects of research literature, tests and measurements, and so forth. Hopefully, you may have even decided that statistics is not impossibly difficult and may be considering taking a more advanced and conventional statistics course. Let me be the first to encourage you and to congratulate you on your sound thinking. But to be sure you really understand what you have learned, please try Module VII.

**#19.** Statistics can be a useful ________________ to organize, describe and make inferences from collections of numberical data.

**#20.** If a student had a Scholastic Aptitude Test (SAT one of the CEEB tests) score of 800, then the evaluation (value judgment) of the score would be that the student had a(an) ________________________________ chance of succeeding in college.

See Appendix A for the answers.

# Part 2

# CONCEPTUALLY APPLYING STATISTICS

MODULE **VII**

# SELECTING AND INTERPRETING STATISTICAL PROCEDURES

Because statistics is essentially a *tool* to enable a person to organize, describe, and perhaps draw inferences from collections of numerical data, this module is included to further help in transferring what you have learned to new situations. Choosing the correct statistical procedure for a specific situation and/or being able to critically evaluate whether the appropriate statistical analysis was used in a particular research study is the supreme test of how well you really understand the preceding material and how useful this information will be for you in the future.

The simplified scrambled-test programmed format simulates the problem solving approach that you will use in choosing, or critically evaluating someone else's choice, of a particular statistical procedure for a specific purpose. Mastery of this material will prepare you for studying and understanding more advanced material such as research methodology books, measurement and evaluation books, encyclopedias of research, handbooks on research or testing, and other major source books for research in education, counseling, psychology, school psychology, health service areas, business, and so forth. This module may become a reference resource to aid in choosing and/or evaluating the choice of statistical techniques for various applications in research studies.

This approach is not difficult and many people find it to be more enjoyable than the usual didactic approach used in the preceding material and in most other introductory statistics textbooks. A hypothetical problem requiring the reader to choose the appropriate statistical procedure is presented. To aid the reader in making a choice three alternative options are presented in a multiple-choice format (Choice A, B, or C). When you make an initial choice of which of the statistical procedures you believe is correct, you then check that choice (Note 38). If you have chosen the "correct answer," this correct choice will be confirmed and you will be presented with another hypothetical multiple-choice problem. If you have chosen an "incorrect answer," this will be confirmed and you will be given additional information to help you understand why another answer would be better. Then you are directed to go back to the original problem, reread and reconsider the problem, and then make a new choice from the remaining options. If you are not entirely sure why the "correct answer" is indeed correct, a good procedure would be for you to review the earlier section of this book that contains information applicable to the relevant specific statistical procedure. A note following each choice indicates the appropriate section(s) for you to review. Reading the explanation provided for each of the three possible answers also may help you understand better the selection of the appropriate statistical procedure.

This is a rather long and challenging module so you may wish to break your study of the material into several different study sessions. Now consider the following (first) problem.

---

**Note 38:** Fictitious names are used in all of the problems in Module VII.

# Problem 1

# TABULAR DESCRIPTION OF STATISTICAL RESULTS

Angus McNew, an educational psychology professor, has prepared a new final examination over the entire course content for his undergraduate classes in learning. The examination contains seventy-five (75) multiple choice questions and was designed to be a norm referenced test that discriminates between those students who have learned most of the material and those students who have learned less of the material. Professor McNew administers the test to one class and collects the test booklets and answer sheets. He then scores each of the answer sheets and arranges the answer sheets in rank order from highest score to lowest score (makes an array of the data) and computes the mean, the median, the mode, and the standard deviation for the scores. Professor McNew wishes to communicate these results to the entire class, but wishes to keep the scores of individual students confidential.

Which of the following statistical procedures will allow him to communicate the general results of this testing but keep the information about individual student scores confidential?

A. construct a circle (pie) graph

B. construct a contingency table

C. construct a frequency distribution

**Answer: A. construct a circle (pie) graph**

Although graphic representations can communicate an overall idea view of scores, they are more appropriately used to demonstrate the general shape of the distribution of scores or the proportions of the group falling into different categories. Circle (pie) graphs are commonly used to show things like the various proportions of the federal tax dollar spent on defense, paying interest of the national debt, and so forth. While a graphic representation might be useful for some purposes, it would not be the appropriate choice for the specific purpose specified in the problem for Professor McNew. Turn back to the original problem and after restudying the problem make a choice between the two remaining options.

**Note:** Graphic presentations are covered in Module I

**Answer: B. construct a contingency table**

Although this is not the best choice it is also not the poorest choice. A tabular presentation is definitely indicated. Tabular presentations do provide a means by which a great deal of data concerning test scores may be communicated without revealing the identity of the person attaining each score. However, contingency tables summarize enumeration (discrete) data which deals with differences in kind or category. The test scores in this problem are measurement (continuous) data. If the data had represented differences in the number (frequency) of individuals who were married, divorced, and so forth, then a contingency table would have been appropriate.

Turn back to the original problem and after restudying the problem choose option C: Constructing a frequency distribution.

**Note:** Tabular presentations are covered in Module I.

**Answer: C. construct a frequency distribution**

Good, you have chosen the best answer to the problem. Large amounts of measurement (continuous) data can be summarized by constructing a frequency distribution and the identity of the individual who attained each score is not revealed. Each student can look at his/her score and compare that score with the other scores.

## Problem 2

# COMPARING DATA BY CONSTRUCTING A GRAPH

Now, turn your attention to a new problem. A psychologist, Selma Zonic, is in the process of constructing and validating a new I.Q. test. She has administered her new instrument to a very large national stratified (on the basis of such things as sex, socioeconomic status, geographic location, and so forth) sample of subjects. Because the belief is that intelligence is normally distributed in the infinite population of human being, she wished to compare the distribution of I.Q. scores for this sample with the hypothesized normal distribution of I.Q. scores.

Which of the following statistical procedures will allow Ms. Zonic to make the most direct comparison of the distribution of I.Q. scores for her sample with the hypothetical normal distribution of the I.Q. scores?

A. construct a histogram

B. construct a frequency polygon

C. construct a smoothed frequency polygon

**Answer: A. construct a histogram**

Although a histogram would allow a person to compare the shape of the distribution of scores for the standardization sample with the hypothetical normal distribution, it would not be the most appropriate graphic distribution to construct. The bar graph nature of a histogram represents the distribution in a stair step fashion which is somewhat different than the smoothed line of the hypothesized normal distribution. It would be better to use a smoothed line to represent both of the distributions which are to be compared. Admittedly the histogram (bar graph) is relatively easy to construct but consistency in the nature of the representations of the distributions to be compared is worth a little extra effort.

Turn back to the original problem and after restudying the problem make a choice between the two remaining options. The above information should make the choice an easy one.

**Note:** Graphic presentations are covered in Module I.

**Answer: B. construct a frequency polygon**

This was not a bad choice, but it was not the best choice. A frequency polygon (line graph) does avoid the stair step effect of a histogram (bar graph) but joining the frequency mid-points of different levels tends to result in a line that shows peaks and valleys in the graphic representation. The smoothed line of the hypothetical normal curve was constructed using an averaging process that smooths out the peaks and fills in the valleys. The rationale for using this averaging process concerns what probably would result if it were possible to assess the infinite population. Therefore,the most appropriate choice would be to use this same averaging process to construct the graphic representation of the scorings in Ms. Zonic's standardization sample.

Turn back to the original problem and choose another option. Remember the information above when you make that choice.

**Note:** Graphic representations are covered in Module I.

**Answer: C. construct a smoothed frequency polygon**

Correct, you have chosen the most appropriate answer. The smoothed frequency polygon is constructed in the same fashion as the hypothetical normal curve was constructed. The rationale for using the smoothed frequency polygon focuses on the idea of what would probably happen (probability) if these data were more numerous and complete. Comparing two smoothed frequency polygons makes for a more direct comparison than comparing a smoothed frequency polygon with either a bar graph or a simple line graph.

## Problem 3

# SELECTING NORMS FOR COMPARISON PURPOSE

Now a new problem for you to consider. William Wronka, a middle school principal realizes that the standardized achievement test battery used in his school does not include a social studies test. Mr. Wronka is planning to do a comprehensive curriculum evaluation in his school and wishes to identify curricular areas where the students are achieving relatively well or poorly compared to their achievement in the other curricular areas. An examination of available standardized social studies achievement tests reveals that some of the tests report only national percentile norms, some report only standard scores, and some provide only regional percentile norms.

Which norms will be most helpful to Mr. Wronka in comparing the relative strength and weakness his students achieve in the various tested curricular areas?

A. national percentile norms

B. standard score norms

C. regional percentile norms

**Note:** Interpreting scores is presented in Module VI.

**Answer: A. national percentile norms**

National percentile norms provide information for interpreting how well a student or how well a group of students have achieved relative to the national sample. However, they are not particularly helpful in identifying relative achievement strengths and weaknesses in different curricular areas within a particular school. In addition percentile norms provide uneven measurement units along the continuum which contributes to error of interpretation of such scores. Standard scores provide even measurement units along the continuum and are relatively easy to convert to a common standard score format (e.g., T-scores, with a mean of 50 and a standard deviation of 10) so that relative strengths and weakness among the mean scores for a group on several different tests may be clearly indicated.

Turn back to the original problem and after restudying the problem make another choice. The discussion above should indicate the most logical choice.

**Note:** Interpreting scores is presented in Module VI.

**Answer: B. standard score norms**

Correct, standard score norms allow you to compare the scores on unlike measures to determine relatively strong or weak performance of the group in question on the various tests. If the scores on each of the various achievement areas is reported as T-scores (mean equals 50 and the standard deviation equals 10), then Mr. Wronka could determine easily that the mean T-score of 40 on the social studies test is considerably lower than a mean T-score of 60 on a mathematics achievement test and a mean T-score of 70 on an English achievement test.

**Answer: C. regional percentile norms**

Regional norms seem to be potentially meaningful to Mr. Wronka, but percentiles are not particularly helpful in clearly identifying relative achievement strengths and weaknesses in different curricular areas within a particular school. In addition percentile norms provide uneven measurement units along the continuum which contribute to error of interpretation of such scores. Standard scores provide even measurement units along the continuum and are relatively easy to convert to a common standard score format (e.g., T-scores, with a mean of 50 and a standard deviation of 10) so that relative strengths and weaknesses among the mean scores for a group on several different tests may be clearly indicated.

Turn back to the original problem and after restudying the problem make another choice. The discussion above should indicate the most logical choice.

**Note:** Interpreting scores is presented in Module VI.

## Problem 4

# DIFFERENTIATING AMONG STANDARD SCORES

Are you ready to consider another problem? Professor Elizabeth Counts made the following statement while lecturing about standard scores. "Deviation IQs, T-scores, and stanines are all forms of standard scores, but one of them is considerably different from the other two." Which of the following is "considerably different" from the other two?"

A. deviation IQs

B. T-scores

C. stanines

**Note:** Module VI deals with interpreting scores and contains a figure showing standard scores and percentile ranks for the normal curve.

**Answer: A. deviation I.Q.s**

Sorry, but this was not the best choice. Although I.Q.s represent a special kind of test, the standard score represented is a very typical kind of score and is like z scores, σ scores, CEEB scores, and so forth. All of these standard scores indicate the mean as a starting midpoint, with divisions of the distribution indicatd in terms of plus or minus full standard deviation units (e.g., a plus one standard deviation or a minus two standard deviations). Six standard deviation units encompass almost the entire distribution for deviation I.Q.s and T-scores.

Turn back to the original problem and choose another option. What do you remember about the positioning of the mean and the width of each unit for stanines.

**Note:** Interpreting scores is presented in Module VI.

**Answer: B. T-scores**

Although T-scores seem to be a special kind of score not directly tied to any I.Q. or aptitude test, the T-score is quite a typical standard score. The mean of 50 points is the starting midpoint and standard deviation units of 10 points are used to demonstrate above or below average performance (e.g., a score of 60 points indicate performance one standard deviation above the mean performance level and a score of 30 points indicate performance two standard deviations below the mean performance level). Six standard deviation units encompass almost the entire distribution for T-scores and deviation I.Q.s.

Turn back to the original problem and choose another option. What do you remember about the positioning of the mean and the width of each unit for stanines?

**Note:** Interpreting scores is presented in Module VI.

**Answer: C. stanines**

This is a good choice. Stanines are standard scores but are quite different from most other standard scores. Each stanine unit is only one half of a standard deviation in width and the mean of the distribution falls at the center of the fifth stanine. Most standard scores have approximately three standard deviation units above the mean and three standard deviation units below the mean. The mean is the midpoint and each standard deviation unit above or below that midpoint is of the same width as every other standard deviation unit.

## Problem 5

# DETERMINING WHICH STANDARD ERROR TO COMPUTE

Another new problem dealing with the concept of standard errors is outlined as follows. A group of school counselors were discussing the proper interpretation of scores of standardized tests which were commonly used in their school system. One counselor maintained that an I.Q. was a specific score and should be interpreted as a precise measure yielding a specific and exact score. Other counselors in the group countered with the argument that all the standardized tests used in the school system were not perfectly accurate and a band of scores within which the true score (score when no measure error is present) probably falls was more appropriate. The first counselor then commented that a band of scores was an interesting idea but how could such an idea be accomplished. The group was quick to point out that one simply needed to use the standard deviation of the test in question and the reliability coefficient of that test to compute the proper standard error. Which standard error should be computed in this instance?

A. standard error of measurement (Sem)

B. standard error of an estimate (Se est)

C. standard error of a mean (Se mean)

**Note:** Standard errors are presented in Module V.

**Answer: A. standard error of measurement (Sem)**

Excellent, the standard error of measurement is the correct solution for the problem posed. The range of scores within which the true score probably falls (with 68% probability) is determined by considering a plus one standard error of measurement above the obtained score (the score "obtained" when the test is administered and scored) and one standard error of measurement below the obtained score. These two points (scores) mark the upper and lower limits of the 68% confidence interval (range of scores). You may remember that in the formula for the standard error of measurement (Note 39), s is the standard deviation of the test and r is the reliability coefficient of the test in question. If a person wishes to be 95% confident that the true score falls within the range of scores, then a plus and a minus two standard error of measurement from the obtained score provide the upper and lower limits of the 95% confidence interval.

---

**Note 39:** $\text{Sem} = s\sqrt{1.00 - r}$

**Answer: B. standard error of an estimate (Se est)**

Sorry, but this is not the correct choice. The standard error of an estimate is concerned with the relative precision of predicting one score by using a different score. For example, how confidently can you predict scores on a certain standardized achievement test by using a certain I.Q. test. Undoubtedly the prediction will not be perfect and the best that can be hoped for is a range of predicted scores within which one can be 68% confident (or 95% confident) that the achievement tests score will "probably" fall. The question raised concerned the accuracy of the score on one test and the standard error of estimate is concerned with the correlation between two different measures (tests).

Turn back to the original problem and choose one of the remaining options. Remember the problem focuses on the precision of a particular score.

**Note:** Standard errors are presented in Module V.

**Answer: C. standard error of a mean (Se mean)**

No, this is not the standard error that the group of counselors were suggesting could be computed using the standard deviation of the tests and the reliability coefficient of that test. The standard error of a mean is generally used to determine an interval for estimating the accuracy of the mean of a sample (in terms of the measurement of a particular variable) as an accurate representation of the mean of the population from which the sample was drawn. The formula for computing the standard error of the mean does use the standard deviation of the sample but does not use standard deviation of the test nor the reliability coefficient of that test.

Turn back to the original problem and choose one of the remaining options. Look carefully at the statistics to be used in computing the correct standard error.

**Note:** Standard errors are presented in Module V.

## Problem 6

# TESTING AN HYPOTHESIS WITH INFERENTIAL STATISTICS

Now it is time to consider a problem where inferential statistics is used to test an hypothesis. Mr. Theodore Stamper, a high school science teacher, has developed a set of linear programmed materials that cover the same curricular content that his conventional textbook covers. He has been able to divide his students into two equated groups. Group A will use the conventional curriculum materials for one semester and Group B will use the new programmed materials for the same span of time. Mr. Stamper has hypothesized that Group B will score statistically significantly higher than Group A on the final examination at the end of the semester. The final examination is a standardized test that provides standard score norms. Which of the following should Mr. Stamper use to test his hypothesis?

A. Mann-Whitney U Test

B. t-test

C. r, product-moment correlation coefficient

**Note:** Module IV deals with inferential statistical tests.

**Answer: A. Mann-Whitney U Test**

Although you could use the Mann-Whitney U Test it would not be the best choice. The U Test is a non-parametric test and non-parametric tests are generally not as statistically powerful as their parametric counterparts. However, the standard score test results specified in the problem represent interval scale data and no good purpose would be served by converting this data to ordinal scale data. The groups were equated prior to conducting the study so the assumptions of homogeneity of variance and normal distributions probably have been met. Remember that the U Test is available and most appropriately used when these assumptions cannot be met.

Turn back to the original problem and choose a more statistically powerful test that best fits Mr. Stamper's study.

**Answer: B. t-test**

Good, this is the most appropriate choice. The hypothesis is concerned with attempting to determine if statistically significant *differences* in mean achievement were demonstrated. The assumption of interval scale data has been met since the achievement test scores are standard scores. These two groups in the study were equated prior to conducting the study so the assumptions of homogeneity of variance and normal distributions probably have been met. The appropriate analysis of variance technique for testing the mean differences between two groups is the t test. This parametric statistical analysis is more powerful than its non-parametric counterpart the Mann-Whitney U Test and therefore preferred.

**Answer: C. r, product-moment correlation coefficient**

No this is not really an appropriate statistical technique to use, consider the design and hypothesis of this problem. You may remember that bivariate correlation is used when you have one group of subjects and two variable measures for each member of the group. The Pearson product-moment correlation coefficient does call for interval and/or ratio scale data so you were not completely incorrect. Correlational techniques are used to determine if similarities (co-relationship) exist between the variables used and Mr. Stamper was trying to determine if statistically significant differences were demonstrated.

Turn back to the original problem and choose an option that best deals with differences between the mean achievement scores of Group A and Group B.

**Note:** Module III deals with correlation and with nominal, ordinal, interval and ratio scales. Module IV deals with inferential statistics.

## Problem 7

# SELECTING STATISTICAL TESTS FOR FACTORIAL STUDY

When Mr. Stamper found statistically signifiant differences in favor of his programmed materials in this first study he began to wonder if the achievement of high, middle, and low ability males and females would be the same or different? In a second study he set up a factorial study with separate groups of male and female with high, middle, and low ability for both the conventional materials and for the programmed materials. He was able to randomly select and to randomly assign students to the various groups. Which of the following statistical tests would be most appropriate?

A. t-test

B. F test, analysis of variance (ANOVA)

C. F test, analysis of covariance (ANCOVA)

**Note:** Inferential statistical tests are presented in Module IV.

**Answer: A. t-test**

The t-test is not the best answer. When you have only two groups the t-test is appropriate. Although you could do a series of t-tests between groups (two at a time) you would not get a clear indication of any possible interactions between sex, ability, and achievement. There are actually twelve groups in the study outlined so an ANOVA technique for more than two groups is indicated. One of the F tests, analysis of variance or analysis of covariance, is indicated.

Turn back to the original problem and choose one of the two remaining options. Consider carefully that random selection and random assignment were used to achieve what amounts to equated groups.

**Note:** Inferential statistical tests are presented in Module IV.

**Answer: B. F test, analysis of variance (ANOVA)**

You are correct. Because multiple groups, random selection, and random assignment were used to meet the assumptions of homogeneous variance and normal distributions, the use of covariates (statistical control variables) is not really necessary.

**Answer: C. F test, analysis of covariance (ANCOVA)**

This is the second best choice. You recognized (right?) that more than two groups were in the study outlined. Actually twelve groups were indicated, so the t-test, which is appropriate for only two groups, would not be a good choice. You seem to have overlooked the fact that random selection and random assignment were used to meet the assumptions of homogeneous variance and normal distributions. The statistical control gained by using premeasures (covariates) is really not necessary and amounts to over analysis of the data.

Turn back to the original problem and review the problem and then read the answer for the F test-analysis of variance.

**Note:** Inferential statistical tests are presented in Module IV.

## Problem 8

## SELECTING A NONPARAMETRIC STATISTICAL TEST

The next problem was encountered by the register, Dr. Judi Jupiter, of a four year college. Several years earlier the college began a special program for helping economically (and often educationally) disadvantaged students adjust to the rigors of college level academic work. Dr. Jupiter has been assigned the task of determining whether there is a statistically significant difference in the number of regular students and special program students who drop out of the college. The sex (male and female) of the regular and special students is also to be taken into account. She realizes that these data from the records are frequency data (frequency count of how many students drop out) relating to the dichotomy of drop-out or stay-in. Which of the following statistical procedures would be the most appropriate choice for analyzing these data.

A. t-test

B. Mann-Whitney U Test

C. Chi-square Test

**Note:** Inferential statistical tests are presented in Module IV.

**Answer: A. t-test**

Wrong. The t-test is applied when certain assumptions are met. The scale data should be interval or ratio data and observed frequency counts do not meet this assumption. This problem calls for a nonparametric statistical test and the t-test is a parametric statistical test. The other two choices are nonparametric statistical tests. One of them uses rank order data while the other uses frequency count data.

Go back to the original problem and choose the option you believe uses frequency count data.

**Note:** Inferential statistical tests are presented in Module IV.

**Answer: B. Mann-Whitney U Test**

You did choose a nonparametric test, but not the correct one. The Mann-Whitney U Test uses rank order data and these data indicated in the problem are frequency count data. In addition the frequency count data deal with a dichotomy, drop-out versus stay-in. When you studied nonparametric statistical tests, which one used frequency count data which was related to a dichotomy?

Turn back to the original problem and choose another option. Remember the new option should specify a nonparametric test that deals with frequency count data related to a dichotomy.

**Note:** Inferential statistical tests are presented in Module IV.

**Answer: C. Chi-square Test**

Very good, you have chosen the best statistical technique for Dr. Juniper to apply in analyzing the data in this problem. The chi-square Test is designed to determine if statistically significant differences exist between the actual (or observed) frequencies expected to fall in each category. Frequency count data (especially as related to a dichotomy such as drop-out versus stay-in) are consistent with the chi-square analysis while rank order data fits the Mann-Whitney U Test and interval data are required for the t-test.

## Problem 9

# TESTING STATISTICAL DIFFERENCES AMONG NON-EQUATED GROUPS

A more complex research design was used by Alfred Buchanan, an educational researcher, in studying the relationship between three different methods of teaching reading to second grade pupils and reading achievement as measured by an achievement test reporting standard score norms. Mr. Buchanan wished also to investigate whether sex (male and female) and socioeconomic status moderated the effects of the different methods of teaching reading on reading achievement. In order to obtain enough second grade pupils he had to use intact classes from several different schools. So random selection of pupils for each of his groups was not possible and he could not assume that the various groups were equated. He knew that I.Q. scores and first grade reading achievement test scores were available for all of the pupils in the different schools. In order to analyze these data to determine if statistically significant differences in achievement occurred between methods of teaching reading, sex, and socioeconomic status, he should use which of the following statistical tests?

A. F test, analysis of covariance (ANCOVA)

B. F test, analysis of variance (ANOVA)

C. R, multivariate correlation coefficient

**Note:** Inferential statistical tests are presented in Module IV.

Correlation is presented in Module III.

**Answer: A. F test, analysis of covariance (ANCOVA)**

Yes, this is the best answer to the problem facing Alfred Buchanan. He is interested in mean differences in reading achievement between the various groups. The multivariate correlation coefficient (R) is basically concerned with similarities rather than differences. The multivariate correlation coefficient (multiple regression coefficient) will be used in the analysis of covariance to equate the probable influence of I.Q. and first grade reading achievement on second grade reading achievement before the F test is done. However, alone the R Test is not the best answer. The F test analysis of variance would have been appropriate if all of the groups had been equated but is not the best choice for the intact groups outlined in the problem.

**Answer: B. F test, analysis of variance (ANOVA)**

This would have been the best answer if the groups had been equated groups. If random selection of pupils for each of the groups had been possible so that the influences of I.Q. and past achievement could have been assumed to be controlled, then the F test for mean differences in achievement between the various groups would have been appropriate. What statistical technique uses premeasures as statistical controls before the significance of differences is computed?

Turn back to the original problem and choose another option. If you are confused by this problem, read both options and review the sections noted below.

**Note:** Inferential statistical tests are presented in Module IV.

Correlation is presented in Module III.

**Answer: C. R, multivariate correlation coefficient**

Essentially any correlation technique is concerned with similarity rather than differences, so this is not the best answer. The R statistic is used in analysis of covariance to equate the probable effect of premeasures (covariates) on the dependent variable. However, this is only a preliminary step leading to the F Test between the mean achievement of the various groups.

Do not be discouraged, because this is a very difficult question. Review the sections noted below and then turn back to the original problem and choose another option.

**Note:** Inferential statistics are presented in Module IV. Correlation is presented in Module III.

## Problem 10

# DECIDING TEST FOR NON-EQUATED GROUPS WITH ORDINAL SCALE DATA

Here is a new problem that is simpler in nature. Miss Monica Harmony, a fifth grade teacher, attended a teacher's convention and learned of a new way of teaching spelling. Because the upper elementary grades are departmentalized, Miss Harmony teaches spelling to two different groups of students. She decides to use her conventional way of teaching spelling with one group and the new approach to the other group. At the end of six weeks she plans to administer her own teacher made spelling test to both groups to determine whether there is a statistically significant difference in the achievement of the two groups. Remember that the two groups are intact rather than equated groups and that her teacher made test scores represent only ordinal scale data. So she should decide to use which of the following statistical tests?

A. chi-square test

B. rho, rank order correlation coefficient

C. Mann-Whitney U Test

**Note:** Inferential statistical tests are presented in Module IV.

Correlation is presented in Module III.

**Answer: A. chi-square test**

The chi-square test does deal with differences, but with differences in expected and actual (or observed) frequencies. The data in Miss Harmony's study is ordinal data. If the data were interval scale data and assumptions of normal distributions and homogeneous variance could be met, then she should use a t test. Do you remember what test is often referred to as the nonparametric counterpart of the t-test?

Turn back to the original problem and choose the option you believe is the nonparametic counterpart of the t-test.

**Note:** Inferential statistical tests are presented in Module IV.

**Answer: B. rho, rank order correlation coefficient**

No, this is the poorest choice for Miss Harmony to use. She is looking for statistically significant differences in spelling achievement (univariate) between two groups. The rank order correlation coefficient does use ordinal data, but for two variables (bivariate) for one group of subjects to determine if a statistically significant similarity (corelationship) exists.

Turn back to the original problem and choose one of the two remaining options. Both remaining options deal with tests of difference but only one option uses the ordinal scale data described in the problem.

**Note:** Inferential statistical tests are presented in Module IV.

**Answer: C. Mann-Whitney U Test**

Very good, you have chosen the appropriate statistical test for Miss Harmony's problem. The Mann-Whitney U Test uses rank order data (ordinal scale data) and is appropriate when the assumptions of homogeneous variance and normal distributions cannot be met. If the spelling test data had been interval scale data and the two groups had been equated groups, then the t-test would have been appropriate. But for Miss Harmony a problem is outlined where these assumptions cannot necessarily be met so the nonparametric counterpart of the t-test is indicated. The Mann-Whitney U Test is often called the nonparametric counterpart of the t-test.

# Problem 11

## DETERMINING CORRELATION BETWEEN TWO SETS OF DATA

Now consider a problem where correlation is used. A college teacher, Zebulon Lorka, regularly gives a pretest during the first session of his classes. One student asks about the relationship between how quickly people finish the pretest and how they score on the pretest. Mr. Lorka has never systematically gathered the data necessary to compute a correlation coefficient between these two variables. So he admits that he does not know what size positive or negative correlation coefficient would result. However he decides to order the finish of each student by recording a 1 for the first person who finishes, a 2 for the second person who finishes, and so on. The pretest scores are simply the number of questions answered correctly. After the pretest is completed and he has recorded the order of finish for each student he decides to compute the correlation coefficient between the raw scores and the order of finish for that group of students. Which of the following correlations should be computed?

A. r, produce-moment correlation coefficient

B. rho, rank order correlation coefficient

C. R, multivariate correlation coefficient

**Note:** Correlation is presented in Module III. Inferential statistical tests are presented in Module IV.

**Answer: A. r, product-moment correlation coefficient**

You have correctly identified that Mr. Lorka is interested in the relationship between two variables (bivariate correlation coefficient) but you have overlooked the fact that the product-moment correlation coefficient requires interval and/or ratio scale data. Mr. Lorka has ordinal data (order of finish) for how quickly people complete the pretest and ordinal data (raw score = ordinal data) for the test scores. He undoubtedly would rank the highest score one, the next highest score two, and so forth. Remember that tied scores each receive the average (mean) rank for those scores.

Turn back to the original problem and choose another option. Remember that you were correct in recognizing that the problem is concerned with a bivariate relationship.

**Note:** Correlation is presented in Module III.

**Answer: B. rho, rank order correlation coefficient**

Very good. The Spearman rank order correlation coefficient is indeed the correct correlation technique for Mr. Lorka to use. Because two variables (raw score and order of finish) are present, a bivariate correlation coefficient is appropriate and both of the variables are ordinal scale measures. You undoubtedly remember that the Pearson product-moment correlation coefficient and the multivariate correlation coefficient most appropriately require interval and/or ratio scale measures.

**Answer: C. R, multivariate correlation coefficient**

I am afraid you chose the poorest answer of the three. You overlooked two important elements outlined in the problem. First, Mr. Lorka's study is concerned with only two variables (bivariate), order of finish of the test and raw score of the test. Second, in the problem is indicted that the measures used provide only ordinal scale data rather than interval and/or ratio scale data. Although some multivariate correlation techniques permit the use of ordinal data (and even nominal data) the use of interval and/or ratio scale data is usually assumed to be most appropriate.

Turn back to the original problem and choose one of the remaining bivariate correlation procedures. Try to choose the one that uses the kind of scale data outlined in this problem.

**Note:** Correlation is presented in Module III.

## Problem 12

# COMPUTING CORRELATION WITH THREE VARIABLES

The next semester Mr. Lorka decides to expand his correlational study. But this time he plans to measure the variables in his new study differently. Rather than recording the order of finish he plans to use a watch to determine how many minutes and seconds it takes each student to complete the test. He has gathered enough data on the spelling test to generate standard score norms for the test. Because his students have considerable variability in their chronological ages, Mr. Lorka decides to include chronological age as an additional variable. Which of the following correlations should he compute?

A. r, product-moment correlation coefficient

B. rho, rank order correlation coefficient

C. R, multivariate correlation coefficient

**Note:** Correlation is presented in Module III. Inferential statistical tests are presented in Module IV.

**Answer: A. r, product-moment correlation coefficient**

If Mr. Lorka had not added a third variable, chronological age, then this would have been the correct answer. But adding a third variable, chronological age, makes this choice an incorrect answer. Adding a third variable goes beyond bivariate correlation (two variables). You were correct in recognizing that the time it took each student to finish the test provides ratio scale data and the standard score normative scores provide interval scale data. Therefore a parametric correlational technique is appropriate. But r is a bivariate parametric correlational technique.

Turn back to the original problem and choose the option which you believe represents a parametric correlational techniques appropriate when three variables are used.

**Note:** Correlation is presented in Module III. Inferential statistical tests are presented in Module IV.

**Answer: B. rho, rank order correlation coefficient**

This answer is incorrect for two basic reasons. First, the scale is of better quality than ordinal data. The time it took each student to finish the test provides ratio scale data and the standard score normative scores provide interval scale data. Second, rho is used for two variables (bivariate correlation coefficient) and Mr. Lorka added chronological age (another ratio scale data variable). So a parametric correlational technique appropriate for three or more variables is indicated.

Turn back to the original problem and choose the option which you believe requires interval and/or ratio scale data and which is appropriate when three variables are used.

**Note:** Correlation is presented in Module III. Inferential statistical tests are presented in Module IV.

**Answer: C. R, multivariate correlation coefficient**

Correct, Mr. Lorka's new study involves three variables (time it took to finish the test, standard score spelling test results, and chronological age) and is a multivariate problem. The time it took to finish the test provides ratio scale data. The standard score normative scores provide interval scale data. Chronological age provides ratio scale data. The assumption of interval and/or ratio scale data for the multiple variables used in the study has been met, so the parametric multivariate R is the correct choice.

## Problem 13

# DECIDING RELATIONSHIP BETWEEN DICHOTOMOUS AND CONTINUOUS VARIABLES

Are you ready for a problem that deals with special correlational methods that were introduced but not covered in as much detail as product-moment and rank order correlational methods? Mrs. June January has had item analysis computed on her new test. The computer print-out of the item analysis provides a correlation coefficient for each item which is labeled "item validity." She does not understand what data was correlated to provide this "item validity" coefficient so she consults the manual provided to explain the various aspects of the item analysis. Mrs. January reads that the total test score (number of items answered correctly) for each person taking the test is correlated with whether the person passed or failed that particular item. After some review she realizes that the "item validity" coefficient is which of the following?

A. point biserial correlation coefficient

B. phi correlation coefficient

C. tetrachoric correlation coefficient

**Note:** Correlation is presented in Module III.

**Answer: A. point biserial correlation coefficient**

Right on! The "item validity" correlation coefficient is computed using the total test score (continuous variable) for each student and whether the student passed or failed the particular test item in question (dichotomous variable). The point biserial correlation coefficient examines the relationship between a continuous variable and a dichotomous variable. In fact this is one of the most frequently used applications of the biserial correlation coefficient in measurement and evaluation. If you took more than one try at choosing the correct option and/or were forced to turn back to Module III for review, you should not be discouraged. This problem involved material which was not presented in great detail. Perhaps it would be a good idea to read the other two options (if you have not already done so) before trying the next problem.

**Answer: B. phi correlation coefficient**

This is not the correct answer. One of the variables (right versus wrong) is a dichotomous variable but the other variable (total test score) is a continuous variable. The phi correlation coefficient is used when both variables are dichotomous variables. If Mrs. January had wished to correlate sex (female or male) with right versus wrong, then the phi correlation coefficient would have been appropriate.

Turn back to the original problem and choose another option. Remember you are looking for the special correlation method that assesses the relationship between a dichotomous variable and a continuous variable.

**Note:** Correlation is presented in Module III.

**Answer: C. tetrachoric correlation coefficient**

Sorry but this is not the correct answer. The tetrachoric correlation coefficient assesses the relationship between two dichotomized variables. Dichotomized variables that are not true dichotomous variables but must be artificially dichotomized. An example of an artificially dichotomized variable would be high scores on a test (scores at or above the mean score) versus low scores. Passing an item or failing an item represents a dichotomous variable and total test score represents a continuous variable (Note 40).

**Note 40:** You may remember the distinction made between continuous variables and discrete variables in Module I.

Turn back to the original problem and choose another option. Remember you are looking for the special correlation method that assesses the relationship between a dichotomous variable and a continuous variable.

**Note:** Correlation is covered in Module III.

## Problem 14

# RELATING df TO T-TEST AND CHI-SQUARE

Brian Singleton, a graduate student, was reviewing a number of empirical research studies which used t-tests (ANOVA) and chi-square statistical analyses. Brian realized that he did not remember very clearly what he had learned about degrees of freedom (df) for these two different inferential statistical tests. In an attempt to clarify his thinking he wrote down the following statements. Which one of the statements is correct?

A. The larger the number of degrees of freedom for both the t-test and the chi-square test the greater the chance of statistical significance for any set value.

B. The smaller the number of degrees of freedom for the t-test and the larger the number of degrees of freedom for the chi-square test the greater the chance of statistical significance for any set value.

C. The larger the number of degrees of freedom for the t-test and the smaller the number of degrees of freedom for the chi-square test the greater the chance of statistical significance for any set value.

*Note:* Inferential statistical tests are presented in Module IV.

**Answer: A. The larger the number of degrees of freedom for both the t-test and the chi-square test the greater the chance of statistical significance for any set value.**

This statement is not correct. Remember that degrees of freedom for the t-test refers to the number of subjects in the two groups and that as sample size increases the confidence in the statistical test results increases. However for the chi-square analysis degrees of freedom refers to the complexity of the chi-square table and the simpler the chi-square table the greater the confidence in the statistical text result. The formula for compluting degrees of freedom for chi-square is rows minus one times columns minus one.

Turn back to the original problem and choose the option that corresponds to the above information.

**Note:** Inferential statistical tests are presented in Module IV.

**Answer: B. The smaller the number of degrees of freedom for the t-test and the larger the number of degrees of freedom for the chi-square test the greater the chance of statistical significance for any set value.**

This statement is not correct. Degrees of freedom for the t-test and for the chi-square test do differ in terms of size and potential statistical significance. However, this statement is the reverse of what is true. Degrees of freedom for the t-test refers to the number of subjects in the two groups. You probably remember that as sample size (in ANOVA tests) increases the confidence in the statistical test results increases. Degrees of freedom for the chi-square analysis refers to the complexity of the chi-square table. The simpler the chi-square table the greater the confidence in the statistical test results. The formula for computing degrees of freedom for chi-square is rows minus one times columns minus one.

The correct choice should now be clear. Turn back to the original problem and choose the correct option.

**Note:** Inferential statistical tests are presented in Module IV.

**Answer: C. The larger the number of degrees of freedom for the t-test and the smaller the number of degrees of freedom for the chi-square test the greater the chance of statistical significance for any set value.**

This is the correct statement. Degrees of freedom for the t-test and for the chi-square test do differ in terms of size and potential statistical significance. Degrees of freedom for the t-test refers to the number of subjects in the two groups and as sample size (in ANOVA tests) increases the confidence in the statistical test results increases. Degrees of freedom for the chi-square analysis refers to the complexity of the chi-square table. The simpler the chi-square table the greater the confidence in the statistical test results. Remember, the formula for computing degrees of freedom for chi-square is rows minus one times columns minus one.

## Problem 15

# INTERPRETING NEGATIVE CORRELATIONS

Here is a problem about another aspect of statistical significance. Kara Carmel read the following statement in the results section of a research study. The product-moment correlation coefficient of -.48 between I.Q. scores and the Smith Scale of Dishonesty (SSD) was significant beyond the .05 level of statistical significance. Which of the following statements provides the most accurate interpretation of what Ms. Carmel read?

A. High level I.Q. scores tended to be matched with high level SSD scores, middle level I.Q. scores tended to be matched with middle level SSD scores, and low level I.Q. scores tended to be matched with low level SSD scores. The actual magnitude of the true relationship between these two variables is -.48.

B. High level I.Q. scores tended to be matched with low level SSD scores and so on, with low level I.Q. scores tending to be matched with high level SSD scores. The magnitude of the true relationship of these two variables is not necessarily precisely -.48.

C. No relationship between I.Q. scores and SSD scores was demonstrated in this study. The odds are that the null hypothesis of no correlation between the two variables is true.

**Note:** Inferential statistical tests (and information concerning statistical significance) is presented in Module IV. Correlation is presented in Module III.

**Answer: A. High level I.Q. scores tended to be matched with high level SSD scores, middle level I.Q. scores tended to be matched with middle level SSD scores, and low level I.Q. scores tended to be matched with low level SSD scores. The actual magnitude of the true relationship between these two variables is -.48.**

This answer is wrong on two counts. First, the relationship described is a direct relationship and represents a positive correlation rather than a negative correlation. Second, a statistically significant correlation coefficient does not necessarily indicate the actual magnitude of the true relationship between the variable described. The correct interpretation is that the odds are against the true magnitude of the relationship being 0.00. Therefore the null hypothesis (0.00 correlation or no correlation between the variables) is rejected. The -.48 correlation coefficient only provides an estimate of the actual magnitude of the relationship between the two variables.

Turn back to the original problem and choose another option.

**Note:** Inferential statistical tests (and information concerning statistical significance) is presented in Module IV. Correlation is presented in Module III.

**Answer: B. High level I.Q. scores tended to be matched with low level SSD scores and so on, with low level I.Q. scores tending to be matched with high level SSD scores. The magnitude of the true relationship of these two variables is not necessarily precisely -.48.**

This answer is correct. First, the relationship described is an inverse relationship and correctly corresponds with the negative correlation coefficient indicated in the problem. The correct interpretation of a statistically significant correlation coefficient is that the odds are against the true magnitude of the relation between the variables being 0.00. In other words, the null hypothesis of no correlation (r = 0.00) is rejected because it is highly unlikely that the true magnitude of the relationship is 0.00. The reported correlation coefficient of -.48 is only an estimate of the magnitude of the actual magnitude of the relationship between the variables.

**Answer: C. No relationship between I.Q. scores and SSD scores was demonstrated in this study. The odds are that the null hypothesis of no correlation between the two variables is true.**

If you chose this option, you should review the Module covering statistical significance before turning back to the original problem and choosing another option. A negative correlation does represent a relationship, even though it is an inverse relationship. Any correlation coefficient that is statistically significant is an indication that the odds are definitely against the magnitude of the true relationship between the variables being 0.00 (no correlation) and involves rejecting (not accepting) the null hypothesis. Please review and then turn back to the original problem and choose another option.

**Note:** Inferential statistical tests (and information concerning statistical significance) is presented in Module IV. Correlation is presented in Module III.

## Problem 16

# INTERPRETING RELIABILITY COEFFICIENT

Wally Witchma, a counselor at Blockholtz High School was appointed to a special committee to evaluate the high school testing program. In carrying out his duties he reviewed a number of tests. The technical manual of one test indicated that the "Spearman-Brown" reliability coefficient for that particular test was .81. He also noted that rather strict time limits were imposed for the completion of several sections of the test. Mr. Witchma correctly prepared which of the following interpretative statements for the other members of the committee?

A. A test-retest reliability coefficient of .81 was reported. This approach to estimating the reliability of the test was appropriate.

B. An adjusted split-half reliability coefficient of .81 was reported. However this approach to estimating reliability is suspect in this instance because speed of responding is a significant aspect of this test.

C. An adjusted split-half reliability coefficient of .81 was reported. This approach to estimating the reliability of this test was appropriate.

**Note:** Reliability and validity are presented in Module V.

**Answer: A. A test-retest reliability coefficient of .81 was reported. This approach to estimating the reliability of the test was appropriate.**

This statement is not correct. When a "Spearman-Brown" reliability coefficient is reported the reader should realize that the split-half technique (one of the internal consistency approaches) was used. The split-half approach provides an underestimate of the true reliability of the test and must be adjusted using the Spearman-Brown "prophecy formula." You may remember the generalization that the more items in a test the higher the reliability. Because splitting the test in half represents correlating two test which are only half as long as the test in question, the split-half coefficient is an underestimate. Since the type of approach to estimating the reliability of the test is incorrectly identified, the second part of the statement is not appropriate.

Turn back to the original problem and choose another option.

**Note:** Reliability and validity are presented in Module V.

**Answer: B. An adjusted split-half reliability coefficient of .81 was reported. However, this approach to estimating reliability is suspect in this instance because speed of responding is a significant aspect of this test.**

Both parts of this statement are accurate and this is the best interpretation for Mr. Witchma to communicate to the other members of the committee. When a "Spearman-Brown" reliability coefficient is reported, then you can conclude that the split-half technique (one of the internal consistency approaches) was used. The split-half approach provides an underestimate of the true reliability of the test and must be adjusted using the Spearman-Brown "prophecy formula." You probably remember the generalization that the more items in a test the higher the reliability. Because splitting the test in half represents correlating two tests which are only one-half as long as the test in question, the split-half coefficient is an underestimate. You also may remember that the internal consistency approaches to reliability are appropriate only for power tests and inappropriate for speeded tests. Therefore, Mr. Witchma is correct in warning the other committee members that the reported reliability coefficient is suspect.

**Answer: C. An adjusted split-half reliability coefficient of .81 was reported. This approach to estimating the reliability of the test was appropriate.**

This interpretative statement is only half correct. The "Spearman-Brown" reliability coefficient is an adjusted split-half approach to estimating the reliability of the test in question. You probably remember that split-half estimates of the test reliability are underestimates of the true reliability of the test and must be adjusted by using the Spearman-Brown "prophecy formula." However this approach is probably not appropriate. An internal consistency approach to estimating test reliability (including the split-half approach) is inappropriate when speed is a significant discriminating factor in a test. Internal consistency approaches are appropriate for power tests but suspect for speed tests.

Turn back to the original problem and choose another option.

**Note:** Reliability and validity are presented in Module V.

## Problem 17

# UTILIZING NATIONAL TEST VALIDITY AND RELIABILITY FOR LOCAL TESTING

Mr. Witchma realizes that reasonable reliability is necessary in order for a test to be valid, but that a reliable test is not necessarily valid. While he is reviewing various achievement tests for different curricular areas, he realizes that his knowledge of what is actually taught in these areas is extremely limited and that this fact makes it difficult for him to judge the validity of these achievement tests. He should probably make which of the following recommendations?

A. Because construct validity of achievement tests is of prime concern the committee should carefully construct a logical definition of achievement.

B. Because concurrent validity of achievement tests is of prime concern the committee should correlate results of the achievement tests with the results of several other available achievement tests.

C. Because content validity of achievement tests is of prime concern the committee should content analyze the curricular content (in each of the different curricular areas) actually taught so that a comparison may be made with the curricular content of each achievement test considered.

**Note:** Reliability and validity are presented in Module V.

**Answer: A. Because construct validity of achievement tests is of prime concern the committee should carefully construct a logical definition of achievement.**

We would certainly hope that the members of the committee already have learned an operational definition of achievement. We also hope that they realize that construct validity is a major concern for I.Q. and personality tests but of less importance for achievement tests. Remember the validity of an achievement test is primarily concerned with the agreement between what has been taught and the degree to which the test represents an adequate sampling of questions over that content universe. The above recommendation might have some merit but it certainly is not the best recommendation for Mr. Witchma to make.

Turn back to the original problem and choose another option.

**Note:** Reliability and validity are presented in Module V.

**Answer: B. Because concurrent validity of achievement tests is of prime concern the committee should correlate the results of the achievement tests with the results of several other available achievement tests.**

The idea of gathering empirical data and analyzing it is not all bad. However, this recommendation would not be particularly helpful to the committee at this point in their work. Concurrent validity and predictive validity (the two criterion related approaches) are both impressive and important for tests like I.Q. tests but of less importance for achievement tests. The prime concern for achievement tests is the degree to which the sample of questions in the test matches the curricular structure of what was taught. In other words, does the content of the test represent an adequate sample of the content universe of what is actually being taught?

Turn back to the original problem and choose another option.

**Note:** Reliability and validity are presented in Module V.

**Answer: C. Because content validity of achievement tests is of prime concern, the committee should content analyze the curricular content (in each of the different curricular areas) actually taught so that a comparison may be made with the curricular content of each achievement tests considered.**

This is the best recommendation. I am quite sure that the committee will not be thrilled with the amount of work that this would entail. But, content validity is of prime concern for achievement tests and the content analysis process is the only sure way of assessing the match between the curricular content of the test and the content of the curriculum actually taught. Other aspects of content validity, such as reading level of the test, are also important aspects of content validity.

## Problem 18

# COMPARING TWO STANDARDIZED TEST SCORES FROM DIFFERENT TESTS

High school principal, Vivian Vestal was examining the cumulative record of a male student whose parents had requested a conference. Two test scores caught her attention. The first was a WISC-R (Wechsler Intelligence Scale for Children-Revised) I.Q. of 130 and the second was a SAT (Scholastic Aptitude Test, one of the CEEB tests) Quantitative score of 400. She knew that the father of the student was a mechanical engineer who owned and operated his own consulting firm. Ms. Vestal felt that these two scores probably would be discussed during the conference, so she wanted to be sure that she clearly understood the relationship between the scores. After some thought and computation she correctly concluded which of the following?

A. Both the WISC-R score and the SAT quantitative score indicate approximately the same level of performance.

B. The performance level indicated by the WISC-R score is considerably higher than the performance level indicated by the SAT Quantitative score.

C. The performance level indicated by the WISC-R score is considerably lower than the performance level indicated by the SAT Quantitative score.

**Note:** Interpreting scores is presented in Module VI.

**Answer: A. Both the WISC-R score and the SAT Quantitative score indicate approximately the same level of performance.**

We certainly hope that Ms. Vestal did not come to this conclusion. The difference in performance between the two tests is one of about three standard deviations. If one correctly converts each score to z scores, one score is a +2.0 z (plus two standard deviations above the mean score) and one score is a -1.0 z (minus one standard deviation below the mean score). If you have not computed the z scores for each of the tests (WISC-R and SAT Quantitative), you should do so before turning back to the original problem and choosing another option.

**Note:** Interpreting scores is presented in Module VI.

**Answer: B. The performance level indicated by the WISC-R score is considerably higher than the performance level indicated by the SAT Quantitative score.**

We sincerely hope that this is what Ms. Vestal concluded. The performance level indicated by the WISC-R score converts to a z score of +2.0 z (plus two standard deviations above the mean score). The performance level indicated by the SAT Quantitative score converts to a z score of -1.0 z (minus one standard deviation below the mean score). If you haven't done this conversion, you might do it now just to confirm this information. Ms. Vestal also may be correct in anticipating that the student's father may be concerned with these scores. This assumption will be particularly true if the father hopes his son will become an engineer and eventually take over the family business.

**Note:** Interpreting scores is presented in Module VI.

**Answer: C. The performance level indicated by the WISC-R score is considerally lower than the performance level indicated by the SAT Quantitative score.**

Ms. Vestal is correct in recognizing that one test score indicates a considerably higher level of performance than the other test score indicates. However she has made an error in identifying the relative performance level indicated by the two tests. If you have not converted these two test scores to z scores, you should do so now. You should discover that the WISC-R score converts to a +2.0 z (two standard deviations above the mean score) and that the SAT Quantitative score converts to a -1.0 z (one standard deviation below the mean score). This difference of three standard deviations would be interpreted as a considerable difference.

Turn back to the original problem and choose another option based on the information supplied above.

**Note:** Interpreting scores is presented in Module VI.

# APPENDICES

## A. Answers to Questions

## B. Abbreviations Used in Statistics

# ANSWERS TO QUESTIONS

Answers for the questions included in the text are as follows:

1. *acquire (learn initially)*
   *retain (remember)*
   *transfer (apply to new situations)*

2. *tool*
   *organize*
   *describe*
   *inferences.*

3. *enumeration (discrete)*
   *measurement (continuous)*

4. Mean for A is *6,*
   Mean for B is *5.8.*

Distribution A is *not* skewed.
Distribution B is *negatively* skewed.

5. The mean deviation is *1.2.*
   The standard deviation is *1.41.*

6. Closest to 0.00 is *B.*
   *Negative correlation is B.*

7. Nearest to 1.00 is *A.*
   Multivariate correlation is represented by *B.*
   Bivariate correlation is represented by *A* and *C*

8. .49

9. Best for prediction would be answer *B.* $r =$ *—.91.* Answer C is larger than 1.00 so it is not a legitimate correlation coefficient.

   From age 40 to 80 the reaction time would be + or *positive* (as ages increase reaction times also increase).

   Correct correlation for money spent and amount of sales tax would be *r (Pearson product moment)* because both measures are ratio scales (money).

   The correlation between chronological age (2 years old to 70 years old) and reaction time would probably be *curvilinear.* As age increased (earlier ages) reaction times would decrease, but at later ages as age increased the reaction times also would increase.

10. Guards against making Type *I* error.

11. The correct answer is *C.* $F_{3,80df}$ when you have four groups, three experimental groups and one control group. Degrees of freedom for groups is the number of groups minus 1.

12. The correct answer is *D.* $F_{2,50df}$. As the number of subjects in the groups increase the size (magnitude) of the F ratio necessary to be statistically significant decreases.

13. The correct answer is *A.* Validity is not an absolute, it is specific to who is being tested where and when.

14. Answer for A.a. is *2. coefficient of stability.*
Answer for A.b. is *3. coefficient of equivalence.*
Answer for A.c. is *1. coefficient of internal consistency.*
Answer for A.d. is *1. coefficient of internal consistency.*

Answer for B is *increase.* (The generalization is that the more items in a test the greater the reliability.)

15. *Test B* is the correct answer. The lower the Sem, the greater the precision of measurements

16. Use a *criterion* referenced test. You are interested in determining if the child has mastered enough of the content (etc.) so that he/she can cope with the higher level reading textbook.

17. z score of +1.0,
T-score of *60,*
within the *7th* stanine,
and *84%ile.*

18. The odds are *68* out of 100. The ± 1 Sem band is the 68% confidence interval.

19. *tool*

20. The student would have an *excellent* chance of succeeding in college. A score of 800 is a +3 standard deviation, 99.9 %ile equivalent and is the top score possible on the test.

Appendix **B**

# ABBREVIATIONS USED IN STATISTICS

**ANCOVA**—analysis of covariance

**ANOVA**—analysis of variance

**df**—degrees of freedom

**F ratio**—same as F test

**F tests**—ANOVA tests involving three or more groups

**K-R20**—Kuder Richardson Formula 20 (used to estimate the split-half coefficients for a test)

**K-R21**—Kuder Richardson Formula 21 (used to estimate the split-half coefficients for a test)

**r**—correlation coefficient (used for interval and/or ratio data)

**$r_{ranks}$**—correlation coefficient (used for ordinal data—other symbols include rho and p)

**R**—multivariate correlation technique

**Se est**—standard error of estimate

**Sem**—standard error of measurement

**Se mean**—standard error of a mean

**t ratio**—same as t test

**T-score**—standard score with a mean of 50 and a standard deviation of 10

**t-test**—ANOVA tests involving two groups

**Type I error**—rejecting the null hypothesis when it was true

**Type II error**—accepting the null hypothesis when it was false

**U test**—Mann-Whitney U Test, a non-parametric equivalent of the t-test (uses ordinal data)

**$x^2$**—chi-square

**z score**—standard score expressed in standard deviation units above or below the mean score.

# INDEX

# INDEX

## D



## E



## F

## N



## O



## P



## Q



## R

## S

# ABOUT THE AUTHOR

Dr. Miller writes from the position of one who has spent over thirty-three years in professional education as a public school teacher, counselor, principal, college professor, and administrator. As a professor in the Department of Educational Psychology, Ball State University, Dr. Miller teachers classes in research methodology, tests and measurements, psychodiagnosis, and directs the School Psychology I Programs.

Dr. Miller has authored articles and monographs in a wide variety of professional publications ranging from the *Journal of Genetic Psychology* to the *Indiana Teacher.* He is also a member of diverse professional organizations such as American Psychological Association, Indiana Psychological Association, National Association of School Psychologists, Phi Delta Kappa, and National Education Association.

Hastings College in Hastings, Nebraska provided the educational program for his B.A. Northern Colorado University provided the educational experiences that enabled him to earn his M.A. and The University of Nebraska provided the educational program for his Ed.D. in Educational Psychology and Measurement.